Instructor's Guide

Concrete Finishing

Level 1

National Center for Construction Education and Research

Wheels of Learning
Standardized Craft Training

Prentice Hall
Upper Saddle River, New Jersey Columbus, Ohio

Prentice-Hall, Inc.
Upper Saddle River, New Jersey 07458

This information is general in nature and intended for training purposes only. Actual performance of activities described in this manual requires compliance with all applicable operations procedures under the direction of qualified personnel. References in this manual to patented or proprietary devices do not constitute a recommendation for their use.

Printed in the United States of America

12

ISBN: 0-13-010249-0

Prentice-Hall International (UK) Limited, *London*
Prentice-Hall of Australia Pty. Limited, *Sydney*
Prentice-Hall of Canada, Inc., *Toronto*
Prentice-Hall Hispanoamericana, S. A., *Mexico*
Prentice-Hall of India Private Limited, *New Delhi*
Prentice-Hall of Japan, Inc., *Tokyo*
Pearson Education Asia, *Singapore*
Editora Prentice-Hall do Brasil, Ltda., *Rio de Janeiro*

Preface

This volume is one of many in the *Wheels of Learning* craft training program. This program, covering more than 20 standardized craft areas, including all major construction skills, was developed over a period of years by industry and education specialists. Sixteen of the largest construction and maintenance firms in the U.S. committed financial and human resources to the teams that wrote the curricula and planned the national accredited training process. These materials are industry-proven and consist of competency-based textbooks and instructor guides.

The *Wheels of Learning* was developed by the National Center for Construction Education and Research in response to the training needs of the construction and maintenance industries. The NCCER is a nonprofit educational entity affiliated with the University of Florida and supported by the following industry and craft associations:

Partnering Associations
- ABC Texas Gulf Coast Chapter
- American Fire Sprinkler Association
- American Society for Training and Development
- American Vocational Association
- American Welding Society
- Associated Builders and Contractors, Inc.
- Associated General Contractors of America
- Carolinas AGC, Inc.
- Carolinas Electrical Contractors Association
- Construction Industry Institute
- Design-Build Institute of America
- Merit Contractors Association of Canada
- Metal Building Manufacturers Association
- National Association of Minority Contractors
- National Association of Women in Construction
- National Insulation Association
- National Ready Mixed Concrete Association
- National Utility Contractors Association
- National Vocational Technical Honor Society
- Painting and Decorating Contractors of America
- Portland Cement Association
- Steel Erectors Association of America
- University of Florida
- Women Construction Owners and Executives, USA

Some of the features of the *Wheels of Learning* program include:
- A proven record of success over many years of use by industry companies.
- National standardization providing "portability" of learned job skills and educational credits that will be of tremendous value to trainees.
- Recognition: upon successful completion of training with an accredited sponsor, trainees receive an industry-recognized certificate and transcript from the NCCER.
- Each level meets or exceeds Bureau of Apprenticeship and Training (BAT) requirements for related classroom training (CFR 29:29).
- Well illustrated, up-to-date, and practical information. All standardized manuals are reviewed annually in a continuous improvement process.

Acknowledgments

This manual would not exist were it not for the dedication and unselfish energy of those volunteers who served on the Technical Review Committee. A sincere thanks is extended to:

Gene Boeke Jr.
Rick Bohan
Ted Bowman
Bob Calderone
Anne Ellis
James (Jamie) A. Farny
Gus Lorber
Norbert Lovata
John Nehasil
Dennis Phillips
Michael J. Schneider
Gustave Rener
Dan Wolf

Contents

Introduction to Concrete Construction and Finishing

Module 23101

NATIONAL
CENTER FOR
CONSTRUCTION
EDUCATION AND
RESEARCH

INTRODUCTION TO CONCRETE CONSTRUCTION AND FINISHING

OBJECTIVES

Upon completion of this module, the trainee will be able to:

1. Define terms associated with concrete construction.
2. Identify the composition and characteristics of concrete.
3. Identify the uses of concrete as a building material.
4. Identify the effect of craftsmanship on finished concrete.
5. Explain the concrete construction process.
6. Identify site operation work requirements.
7. Explain the career potentials in concrete construction and finishing.

Note to the Instructor

Before teaching this module, you should review the details in this Instructor's Guide for Equipment and Supplies, Testing, and the suggested Teaching Sequence. Be sure to allow ample time to prepare your own training plan or lesson plan and to gather all required equipment and materials.

Required Equipment and Materials

The following are required for instruction using this module:

Equipment
Overhead projector and screen
Whiteboard/chalkboard
Appropriate Personal Protective Equipment

Materials
Trainee Task Module
Small container of cement
Small container of sand
Small container of gravel
Small container of water
Transparencies
Module Examinations
Markers/chalk
Pencils and paper

HOW TO USE THIS INSTRUCTOR'S GUIDE

For each 2½ hour class session in this Instructor's Guide, the basic Presentation Sequence is as follows:

Introduction/Overview

Classroom, and/or Demonstration, and/or Laboratory

Class Break

Classroom, and/or Demonstration, and/or Laboratory

Summary

Suggested time periods for classroom sessions are included throughout this Instructor's Guide. These time periods should be adapted to meet local conditions and training requirements.

Each class session is presented with two columns of information. On the left side of the page, a narrow column provides suggested trainee and instructor actions, icons to call your attention to material, safety, audiovisual, or testing requirements, and space for your notes. The right-hand column provides the outline of the suggested presentation for each class session.

In this Instructor's Guide, the terms classroom, demonstration, and laboratory are defined and used as follows.

Classroom: Sessions are designed for lectures, group discussions, coaching, and additional activities. Trainees should be encouraged to actively participate.

Demonstration: Instructors will demonstrate all procedures before trainees attempt them. Instructors should make sure that trainees can point out all safety procedures during demonstrations to be assured of the proper use of equipment by trainees.

Laboratory: Instructors will facilitate all laboratory activities, coach trainees as they practice the procedures, monitor trainee progress, and provide feedback. The instructor will make sure that safety rules are followed at all times and that protective equipment is worn.

NCCER Standardized Craft Training Programs

The National Center for Construction Education and Research (NCCER) provides a standardized national program of accredited craft training. Key features of the program include instructor certification, competency-based training, and performance testing. The program provides trainees, instructors, and companies with a standard form of recognition through a National Craft Training Registry. The program is described in full in the *Guidelines for Accreditation*, published by the NCCER. For more information on standardized craft training, contact the NCCER at P.O. Box 141104, Gainesville, FL 32614-1104; or call 352-334-0911.

MODULE OVERVIEW

This module introduces the Concrete Finishing trainee to the methods and procedures used in concrete finishing.

Prerequisites

Please see the Course Map. Prior to training with this module, it is recommended that the trainee shall have successfully completed the following modules:

Core Curricula

Safety Considerations

Ensure that the trainees are equipped with Appropriate Personal Protective Equipment.

Teaching Time for This Module

Approximately 10 hours or four sessions of training time are suggested to cover *Introduction to Concrete Construction and Finishing*. The training class session is a suggested 2½ hour time period, which includes at least one break. **You will need to adjust the time required for hands-on activities and testing based on your class size and resources.** All time periods for this module are suggested and you will need to adapt the suggested lesson plan to meet your local conditions.

Suggested Teaching Sequence — Four 2½-Hour Sessions

Adjust your class times based on class size and resources.

Session	Topic	Trainee Module Section(s)
1	Introduction; What is Concrete?; Uses of Concrete	1.0.0 – 3.1.4
2	Uses of Concrete	3.2.0 – 3.2.7
3	Laboratory/Field Trip	3.2.0 – 3.2.7
4	Importance of Craftworkers; Careers in Concrete Construction and Finishing; Work Requirements	4.0.0 – 6.4.0
	Review/Practice Questions	
	Module Examination	

Optional References for Advanced Study

This module is intended to present thorough resources for task training. The following reference works are suggested for both instructors and motivated trainees interested in further study. These are optional materials for continued education rather than for task training.

Concrete Construction and Estimating, Craig Avery, Craftsman Book Company, Carlsbad, CA, 1991.

Concrete Manual, International Conference of Building Officials, Whittier, CA, 1992.

Concrete Parking Lots: Eight Steps to Quality Construction, Videotape, Portland Cement Association, Skokie, IL.

Concrete Technology, Third Edition, George R. White, Delmar Publishers, Inc., Albany, NY, 1991.

Design and Control of Concrete Mixtures, Thirteenth Edition, Portland Cement Association, Skokie, IL, 1988.

Finishing Concrete Flatwork, Videotape 32:00 minutes, Portland Cement Association, Skokie, IL, 1991.

Guide for Concrete Floor and Slab Construction, ACI 302.1R-96, American Concrete Institute, Farmington Hills, MI, 1997.

The Contractor's Guide to Quality Construction, American Concrete Institute, Farmington Hills, MI, 1997.

COURSE MAP

This course map shows all of the modules in the first level of the Concrete Finishing curricula. The suggested training order begins at the bottom and proceeds up. Skill levels increase as a trainee advances on the course map. The training order may be adjusted by the local Training Program Sponsor.

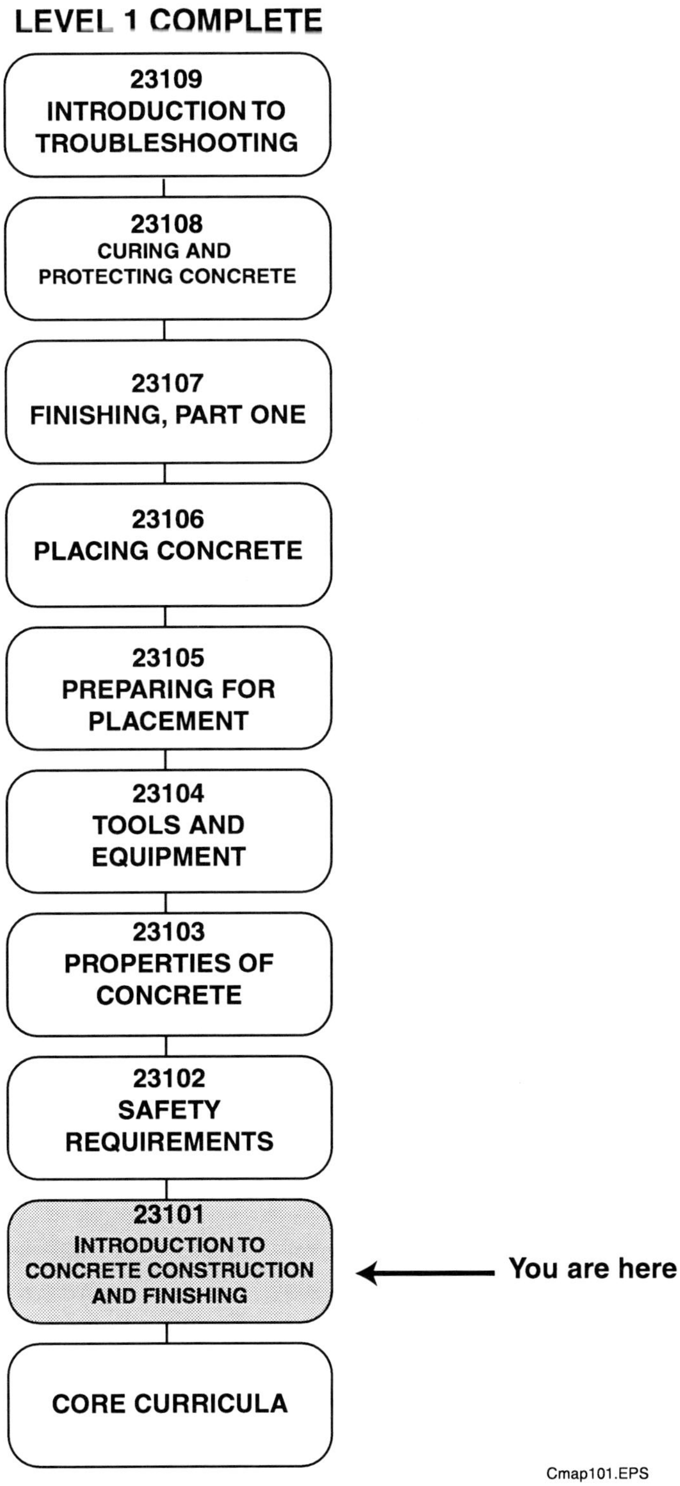

Cmap101.EPS

SESSION 1 OF 4 SESSIONS

Ensure that all trainees have brought their Trainee Modules to class.

Equipment and Materials

Trainee Task Module
Overhead projector and screen
Whiteboard/chalkboard
Markers/chalk
Small container of cement
Small container of sand
Small container of gravel
Small container of water
Transparencies

Time Schedule (150 minutes)

10 Overview
40 Classroom
10 Class Break
40 Classroom
10 Class Break
25 Classroom
15 Summary

Overview (10 minutes)

Show Transparency 1, Course Objectives.

1. Take attendance and perform other administrative tasks.
2. Outline briefly the topics to be covered. Explain to the trainees that they must pass the Module Examination with a grade of at least 70 percent. Explain that there will be no Performance Profile Examination for this module.

Classroom (40 minutes)

1. Review the trade terms introduced in this module and explain them.
2. Show trainees or pass through the class the containers of cement, sand, gravel, and water. Discuss the ingredients of concrete.

Show Transparency 2, Composition of Air-Entrained Concrete (Figure 1, Trainee Module).

3. Tell trainees about the characteristics of cement:
   - Availability
   - Versatility
   - Strength
   - Durability

AUDIO/VISUAL

4. Discuss the history of cement. Tell trainees about experiences which you (or your company) have had using concrete, and how things have changed.

Show Transparency 3, Manufacturing Process for Portland Cement, Quarrying (Figure 2, Trainee Module).

5. Tell trainees about the process of quarrying and blending the raw materials for portland cement.

Show Transparency 4, Manufacturing Process for Portland Cement, Proportioning (Figure 2, Trainee Module).

6. Discuss the processes of proportioning and fine grinding.

Show Transparency 5, Manufacturing Process for Portland Cement, Kiln System (Figure 2, Trainee Module).

7. Tell trainees about the firing process in a modern kiln, and the subsequent manufacture of clinker.

Show Transparency 6, Manufacturing Process for Portland Cement, Finish Grinding (Figure 2, Trainee Module).

8. Discuss the finish grinding of clinker, and its shipping. Note that some cement is packaged rather than shipped in bulk.

9. Tell trainees about pozzolans, air entraining admixtures, and plasticizing agents.

Class Break (10 minutes)

Classroom (40 minutes)

Show Transparency 7, Concrete Strength in Compression and Tension (Figure 3, Trainee Module).

1. Discuss compression and tension as forces acting on concrete. Ask trainees for examples of each force acting on a structural member.

Show Transparency 8, Concrete Beam Failure in Flexure (Figure 4, Trainee Module).

2. Describe load bearing on a slab. Introduce the idea of flexural strength as resistance to both compression and tension.

Show Transparency 9, Reinforced Concrete Beam in Flexure (Figure 5, Trainee Module).

3. Discuss flexural strength.
   - Tell trainees about improvements in flexure with the addition of reinforcing steel.
   - Ask trainees for examples of concrete structural members that would rely on flexural strength.

Show Transparency 10, Example Strength Gain Curves for Concrete (Figure 6, Trainee Module).

4. Review the strength gain curves with trainees. Discuss the relationship between water-cement ratio and speed of strength gain.

5. Review the concepts of prestressed concrete structural members. Identify any local projects containing prestressed members.
6. Review the concepts of post-tensioned concrete structural members. Identify any local projects containing post-tensioned members.
7. Review the concepts of precast concrete structural members. Identify any local projects containing precast members.
8. Review the concept of reinforced concrete and improved flexure. Identify any local projects containing reinforced members.
9. Review the concept of unreinforced, plain concrete. Identify any local projects containing plain concrete members.

Class Break (10 minutes)

Classroom (25 minutes)

1. Discuss and answer questions about slabs on grade.
   - Ask trainees to identify local projects with slabs on grade.
2. Discuss and answer questions about elevated structural elements.
   - Ask trainees to identify local projects with elevated concrete structural elements.
3. Discuss and answer questions about pipes and culverts.
   - Ask trainees to identify local projects with concrete pipes and culverts.
4. Discuss tilt-up construction. Identify any local projects that used tilt-up construction.
5. Discuss custom forming, as shown in *Figure 12* in the Trainee Module.
   - Ask trainees for examples of custom concrete forming in your local area.

Summary (15 minutes)

1. Review the materials covered in this session.
2. Answer trainee questions.
3. Give assignments for for next session.

SESSION 2 OF 4 SESSIONS

Ensure that all trainees have brought their Trainee Modules to class.

Equipment and Materials

Trainee Task Module
Overhead projector and screen
Whiteboard/chalkboard
Markers/chalk
Transparencies

Time Schedule (150 minutes)

5 Overview
40 Classroom
10 Class Break
40 Classroom
10 Class Break
35 Classroom
10 Summary

Show Transparency 1, Course Objectives, and identify the topics to be covered in this session.

Overview (5 minutes)

1. Take attendance and perform other administrative tasks.
2. Outline briefly the topics to be covered.

Classroom (40 minutes)

1. Discuss the beginnings of the construction process:
   - property acquisition
   - contract
   - survey
   - locating utilities
   - setting benchmarks and stakes

Show Transparency 11, Typical Residential Site Plan (Figure 15, Trainee Module).

2. Note the points of interest to concrete workers on the site plan:
   - elevation
   - grade lines
   - measurements/distances
3. Review with trainees the tools for blueprint reading:
   - dimension markings
   - sectional view lines
   - materials symbols

Show Transparency 12, Typical Residential Foundation Plan (Figure 16, Trainee Module).

4. Discuss the residential foundation plan. Identify and discuss these components on the plan:
   - dimensional markings
   - posts
   - sectional views
   - footing details for building and block cross-section
   - footing details for concrete slab
5. Introduce the topic of forming. Remind trainees of the weight and plasticity of fresh concrete. Ask trainees why forms are important in concrete work.

Show Transparency 13, Job-Built Wall Form (Figure 17, Trainee Module).

6. Tell trainees about job-built forms:
   - process of constructing
   - advantages

Show Transparency 14, Job-Built Form for Steps (Figure 18, Trainee Module).

7. Discuss special forming for custom jobs such as stairs-to-fit.

Show Transparency 15, Prefabricated Wall Form (Figure 19, Trainee Module).

8. Tell trainees about prefabricated forms:
   - process of assembly
   - reuse
9. Tell trainees about manufactured forms:
   - special shapes
   - reuse

Class Break (10 minutes)

Classroom (40 minutes)

1. Discuss the process of placing concrete. Tell trainees why placing is important in concrete work. Tell trainees that this is where "hands-on" concrete work begins

Show Transparency 16, Plant Exit (Figure 2, Trainee Module).

2. Describe how concrete gets from the plant to the job site.
3. Tell trainees about moving concrete from the ready-mix truck to the deposit site.
4. Discuss consolidating. Tell trainees why consolidating is important:
   - eliminates voids
   - blends layers together

AUDIO/VISUAL

5. Give trainees a short overview of the finishing process:
    - screeding
    - bullfloating
    - edging
    - jointing
    - floating
    - troweling
    - brooming
6. Discuss screeds.
7. Discuss bullfloats.
8. Discuss highway straightedges.
9. Discuss edging.
10. Discuss power floating.
11. Discuss hand troweling.

Class Break (10 minutes)

Class Session (35 minutes)

Show Transparency 17, Joints In A Concrete Slab (Figure 27, Trainee Module).

1. Tell trainees about cracking in slabs and how joints control cracking.
2. Discuss contraction joints.
3. Discuss isolation joints.
4. Discuss construction joints.
5. Talk about edging.
6. Describe the curing process.

Show Transparency 18, Curing with Sprayed Water To Keep Concrete Moist (Figure 28, Trainee Module).

7. Discuss curing by replacing moisture.
8. Discuss curing by sealing the surface.
9. Discuss curing by accelerating strength gain.
10. Discuss form removal.

Summary (10 minutes)

1. Review the materials covered in class.
2. Answer trainee questions.
3. Give assignments for next session. Ask trainees to begin module review questions.
4. Discuss field trip with trainees. Remind them to bring notebooks and writing tools so they can take notes of their visit.

SESSION 3 OF 4 SESSIONS

Equipment and Materials

Ensure that all trainees have brought writing materials to the site.

Pencils and Paper

Appropriate Personal Protective Equipment

Time Schedule (150 minutes)

15 Overview

120 Site Visit

15 Summary

Overview (15 minutes)

1. Take attendance and perform other administrative tasks.
2. Discuss taking notes and assign specific responsibilities.
3. Impress upon trainees the importance of:
   - remaining alert
   - staying with the group
   - obeying instructions
   - wearing Appropriate Personal Protective Equipment

Ensure that all trainees wear Appropriate Personal Protective Equipment.

Site Visit (120 minutes)

1. On the site tour, point out construction stages and other material covered in the previous two sessions.
2. Answer questions from trainees.

Summary (15 minutes)

1. Review the events and items observed.
2. Give assignments for the next session.
3. Ask trainees to complete Review/Practice Questions for the next session.
4. Remind trainees that the next session will be the last session and that a Module Examination will be given.

SESSION 4 OF 4 SESSIONS

Ensure that all trainees have brought their Trainee Modules to class.

Equipment and Materials

Trainee Task Module
Overhead projector and screen
Whiteboard/chalkboard
Markers/chalk
Transparencies
Module Examinations

Time Schedule (150 minutes)

5 Overview
50 Classroom
10 Class Break
50 Classroom
15 Class Break
30 Module Examination

Show Transparency 1, Course Objectives.

Overview (5 minutes)

1. Take attendance and perform other administrative tasks.
2. Outline briefly the topics to be covered. Explain to the trainees that they must pass the Module Examination with a grade of at least 70 percent.

Classroom (50 minutes)

1. Call on trainees to review the site tour. Cover as many of the work stages as were seen:
   - forming
   - placing
   - consolidating
   - floating
   - jointing and edging
   - troweling
   - brooming
2. Allow trainees to share observations and ask additional questions about material seen.
3. Begin discussion on the importance of concrete craftworkers.
4. Tell trainees how craftworkers are involved in the mixing process.
5. Tell trainees how craftworkers are involved in placing.
6. Tell trainees how craftworkers are involved in curing.

AUDIO/VISUAL

Class Break (10 minutes)

Classroom (50 minutes)

1. Discuss career ladder with trainees:
   - trainee
   - journeyperson
   - foreperson
   - superintendent

Show Transparencies 19 and 20, Craftworker Knowledge.

2. Discuss the skills and knowledge needed for good craftwork.
3. Discuss good craftworker attitudes:
   - dependable
   - team player
   - proud
   - safety-conscious
   - quality-conscious
4. Discuss quality as the goal of craftworkers.
5. Go over review questions with trainees. Answer all unanswered questions and make necessary clarifications.

Class Break (15 minutes)

TESTING

Module Examination (30 minutes)

1. Hand out Module Examinations to all trainees. Tell them they have 30 minutes to complete the exam.
2. Collect and grade all examinations.

MODULE EXAMINATIONS AND ANSWER KEYS

The masters for Module Examinations and Answer Keys are found in the booklet entitled "Concrete Finishing Level One Module Examinations" included in this Instructor's Guide package. You may also reproduce the tests using the Windows-compatible software program which may be purchased separately from Prentice Hall. If you do not have the reproducible masters in your Instructor's Guide package, or would like to order the test software, please contact Prentice Hall Publishers at 1-800-922-0579.

Be sure to keep answer keys and tests in a safe area.

OVERHEAD TRANSPARENCIES

Masters for overhead transparencies are found on the following pages. Transparencies can be made by using clear acetate sheets in a photocopy machine. This can be done at any commercial copy shop, or by following this method on your office machine:

1. Use standard 8½ × 11 clear acetate sheets, which can be obtained at office supply stores, stationery stores, or commercial copying centers.
2. Check the photocopier's manual to make sure it will take the thicker acetate sheets.
3. Remove the paper supply in the photocopying machine and replace it with a supply of acetate sheets.
4. Make photocopies from the masters on the following pages in the usual way.
   The material will be reproduced on the clear acetate sheets for use on any overhead projector.

TRANSPARENCY 1

TASK MODULE 23101, INTRODUCTION TO CONCRETE CONSTRUCTION AND FINISHING

COURSE OBJECTIVES

Upon completion of this module, the trainee will be able to:

1. **Define terms associated with concrete construction.**
2. **Identify the composition and characteristics of concrete.**
3. **Identify the uses of concrete as a building material.**
4. **Identify the effect of craftsmanship on finished concrete.**
5. **Explain the concrete construction process.**
6. **Identify site operation work requirements.**
7. **Explain the career potentials in concrete construction and finishing.**

TRANSPARENCY 2

FIGURE 1

TASK MODULE 23101, INTRODUCTION TO CONCRETE CONSTRUCTION AND FINISHING

COMPOSITION OF AIR-ENTRAINED CONCRETE

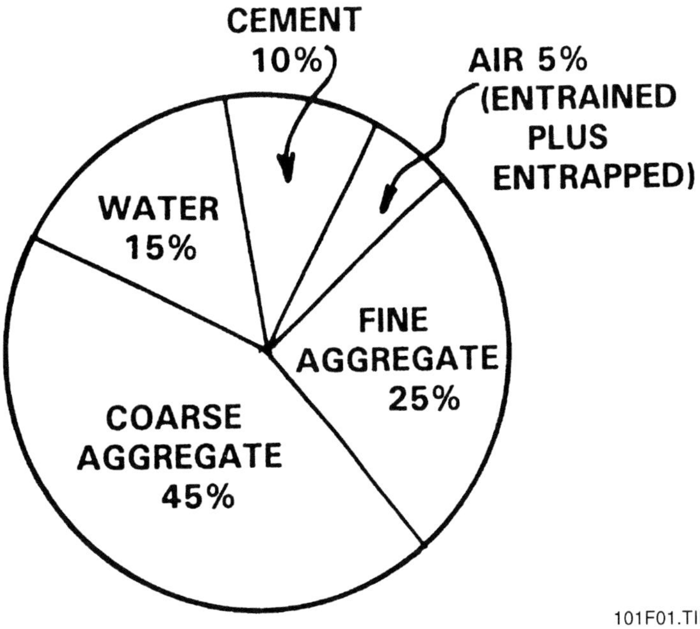

101F01.TIF

TRANSPARENCY 3

FIGURE 2

TASK MODULE 23101, INTRODUCTION TO CONCRETE CONSTRUCTION AND FINISHING

MANUFACTURING PROCESS FOR PORTLAND CEMENT, QUARRYING

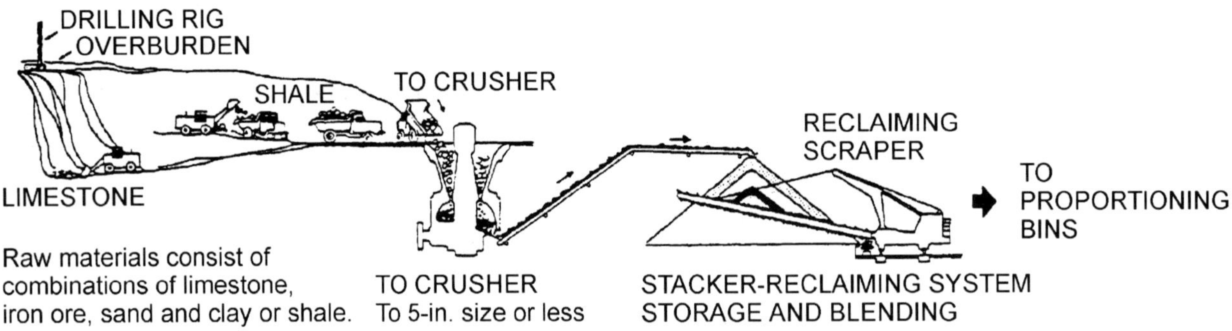

1. Quarrying and blending of raw materials.

TRANSPARENCY 4

FIGURE 2

TASK MODULE 23101, INTRODUCTION TO CONCRETE CONSTRUCTION AND FINISHING

MANUFACTURING PROCESS FOR PORTLAND CEMENT, PROPORTIONING

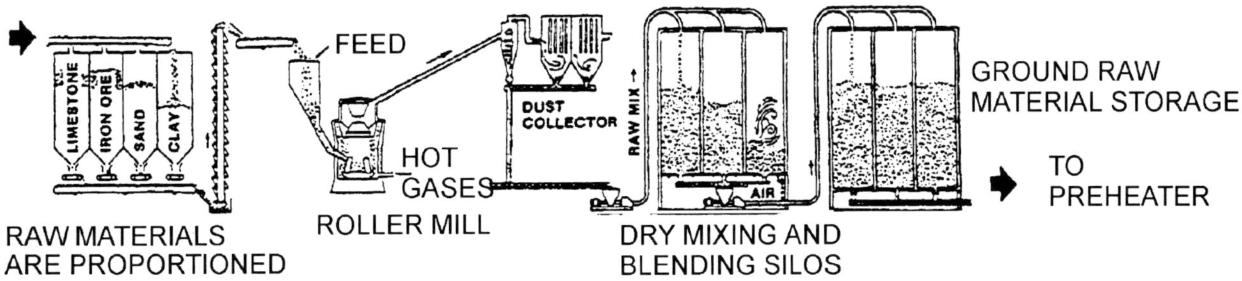

2. Proportioning and fine grinding of raw materials.

TRANSPARENCY 5

FIGURE 2

TASK MODULE 23101, INTRODUCTION TO CONCRETE CONSTRUCTION AND FINISHING

MANUFACTURING PROCESS FOR PORTLAND CEMENT, KILN SYSTEM

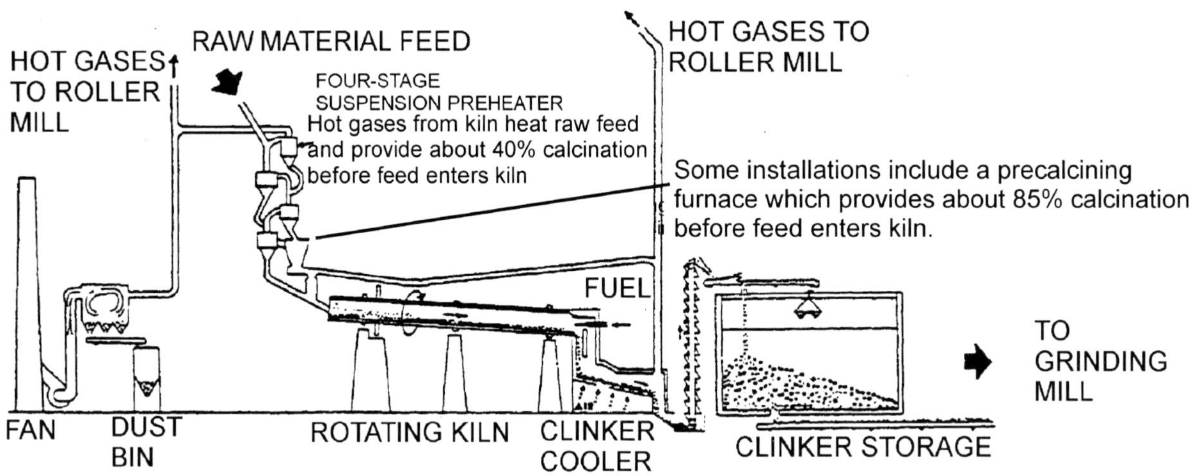

3. Kiln system. Preheating; burning; cooling; and clinker storage.

TRANSPARENCY 6

FIGURE 2

TASK MODULE 23101, INTRODUCTION TO CONCRETE CONSTRUCTION AND FINISHING

MANUFACTURING PROCESS FOR PORTLAND CEMENT, FINISH GRINDING

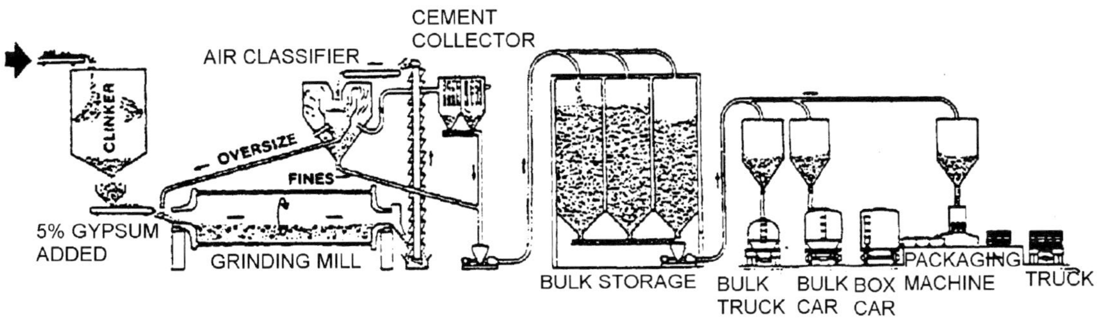

4. Finish grinding and shipping.

101F02.TIF

TRANSPARENCY 7

FIGURE 3

TASK MODULE 23101, INTRODUCTION TO CONCRETE CONSTRUCTION AND FINISHING

CONCRETE STRENGTH IN COMPRESSION AND TENSION

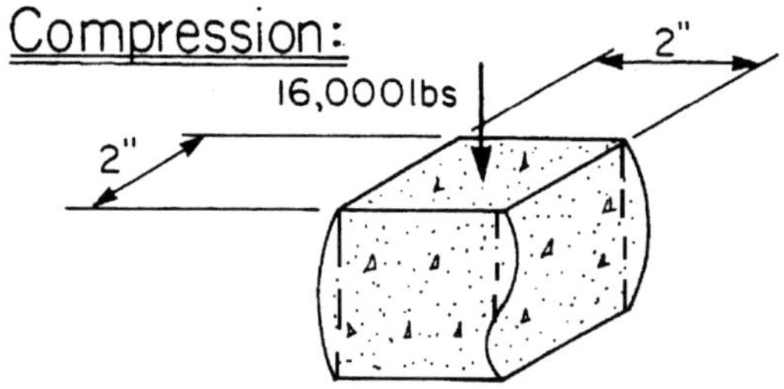

$$f'_c = \frac{16{,}000 \text{ lbs}}{(2'' \times 2'' = 4 \text{ sq.in.})}$$

$$= 4000 \text{ psi}$$

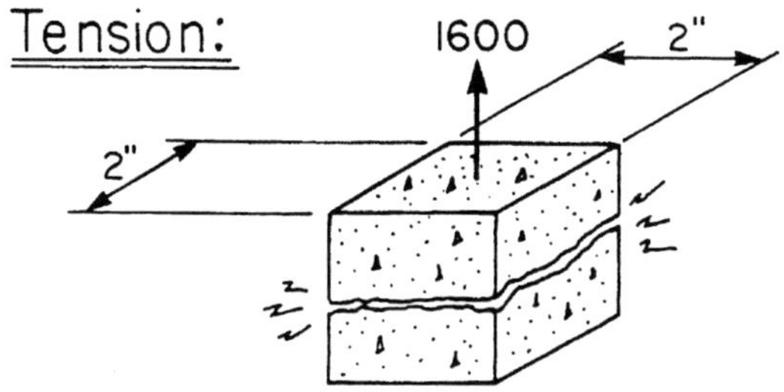

$$f_t = \frac{1600}{(2'' \times 2'' = 4 \text{sq. in.})}$$

$$= 400 \text{psi}$$

101F03IG.TIF

TRANSPARENCY 8

FIGURE 4

TASK MODULE 23101, INTRODUCTION TO CONCRETE CONSTRUCTION AND FINISHING

CONCRETE BEAM FAILURE IN FLEXURE

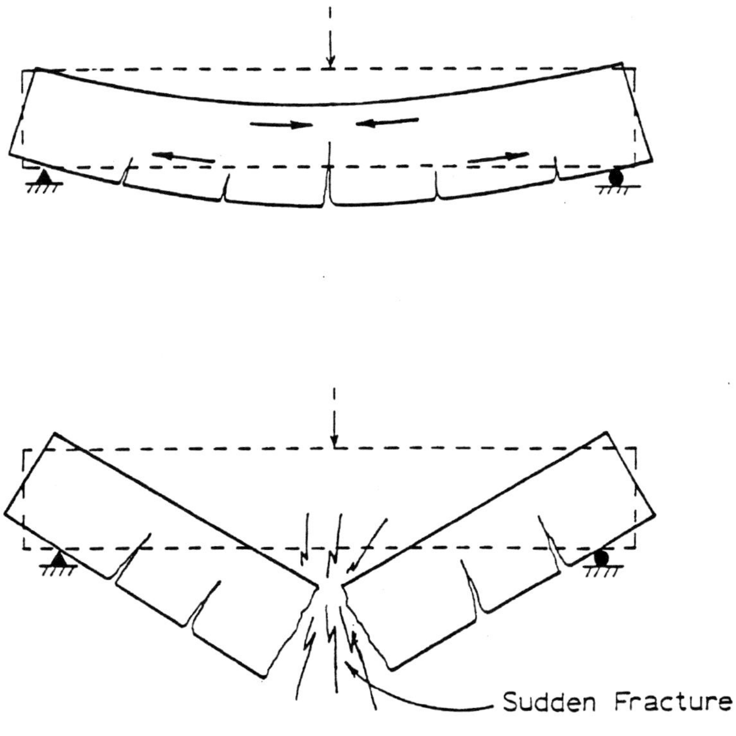

101F04IG.TIF

TRANSPARENCY 9

FIGURE 5

TASK MODULE 23101, INTRODUCTION TO CONCRETE CONSTRUCTION AND FINISHING

REINFORCED CONCRETE BEAM IN FLEXURE

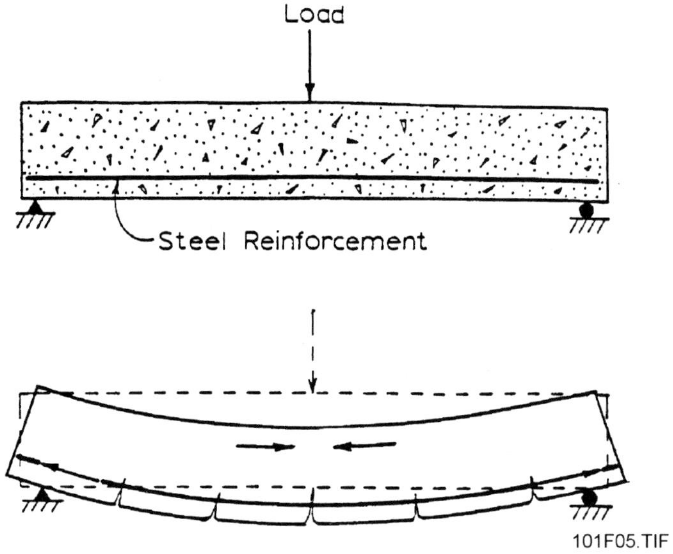

TRANSPARENCY 10

FIGURE 6

TASK MODULE 23101, INTRODUCTION TO CONCRETE CONSTRUCTION AND FINISHING

EXAMPLE STRENGTH GAIN CURVES FOR CONCRETE

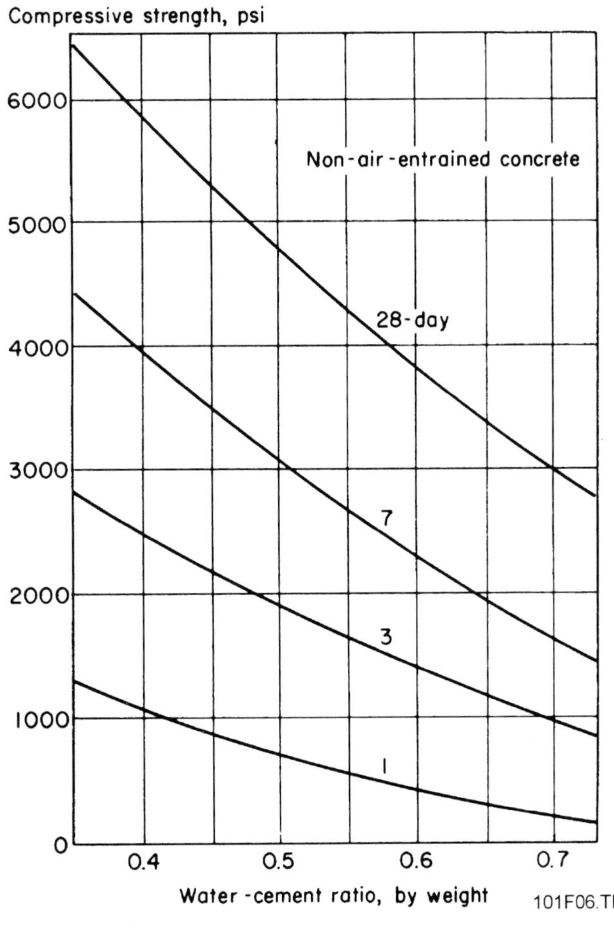

101F06.TIF

TRANSPARENCY 11

FIGURE 15

TASK MODULE 23101, INTRODUCTION TO CONCRETE CONSTRUCTION AND FINISHING

TYPICAL RESIDENTIAL SITE PLAN

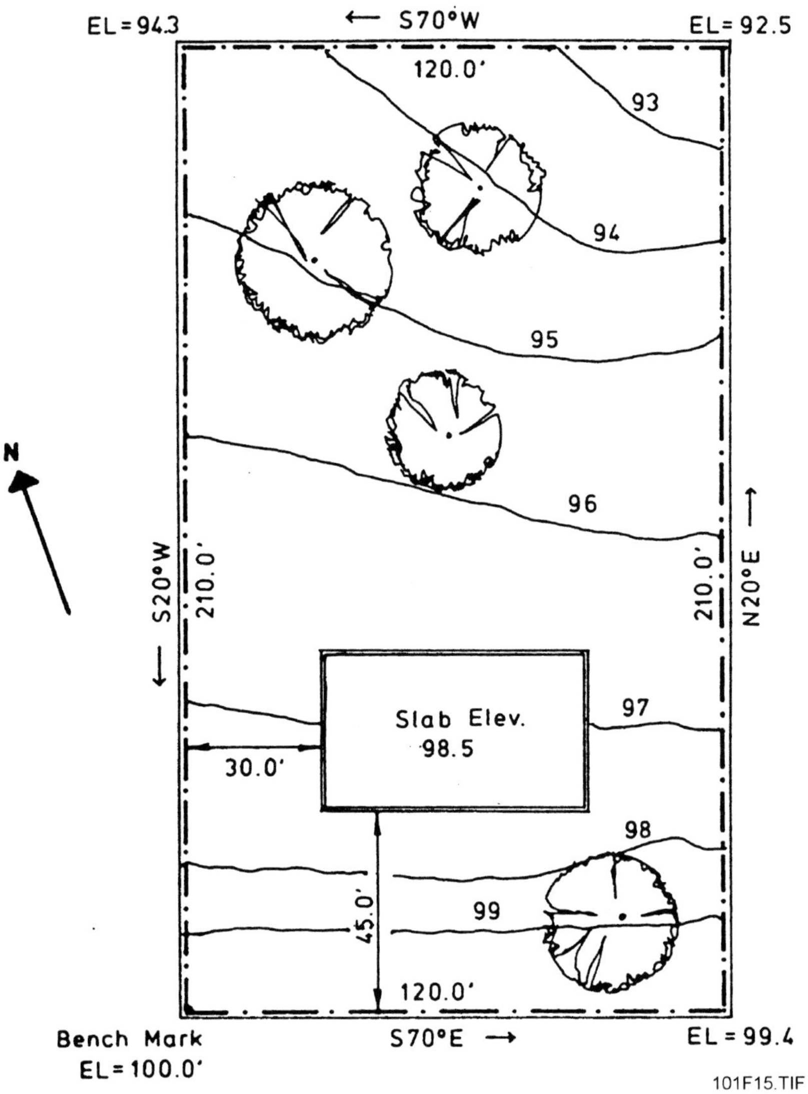

101F15.TIF

TRANSPARENCY 12

FIGURE 16

TASK MODULE 230101, INTRODUCTION TO CONCRETE CONSTRUCTION AND FINISHING

TYPICAL RESIDENTIAL FOUNDATION PLAN

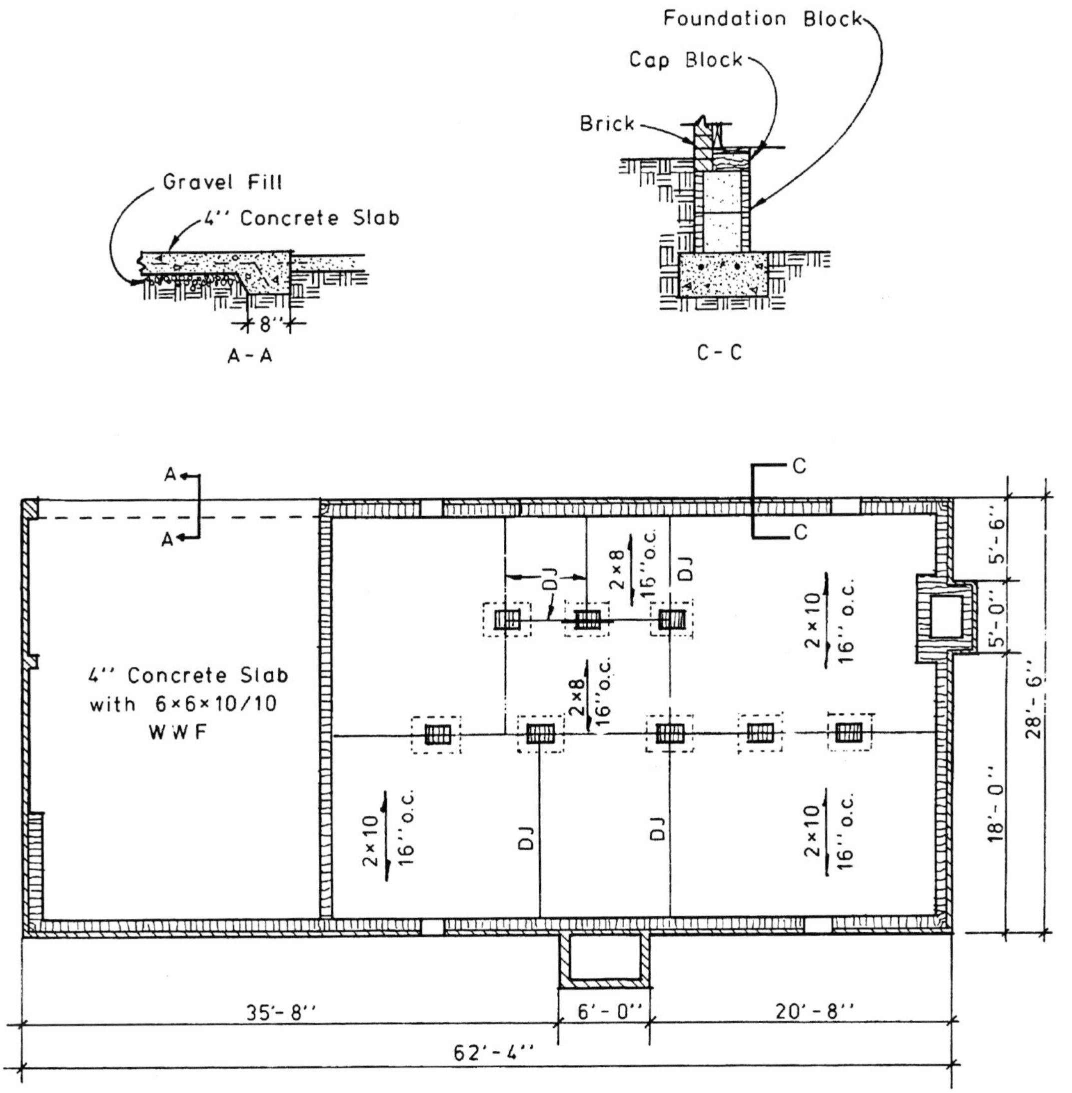

101F16.TIF

TRANSPARENCY 13

FIGURE 17

TASK MODULE 23101, INTRODUCTION TO CONCRETE CONSTRUCTION AND FINISHING

JOB-BUILT WALL FORM

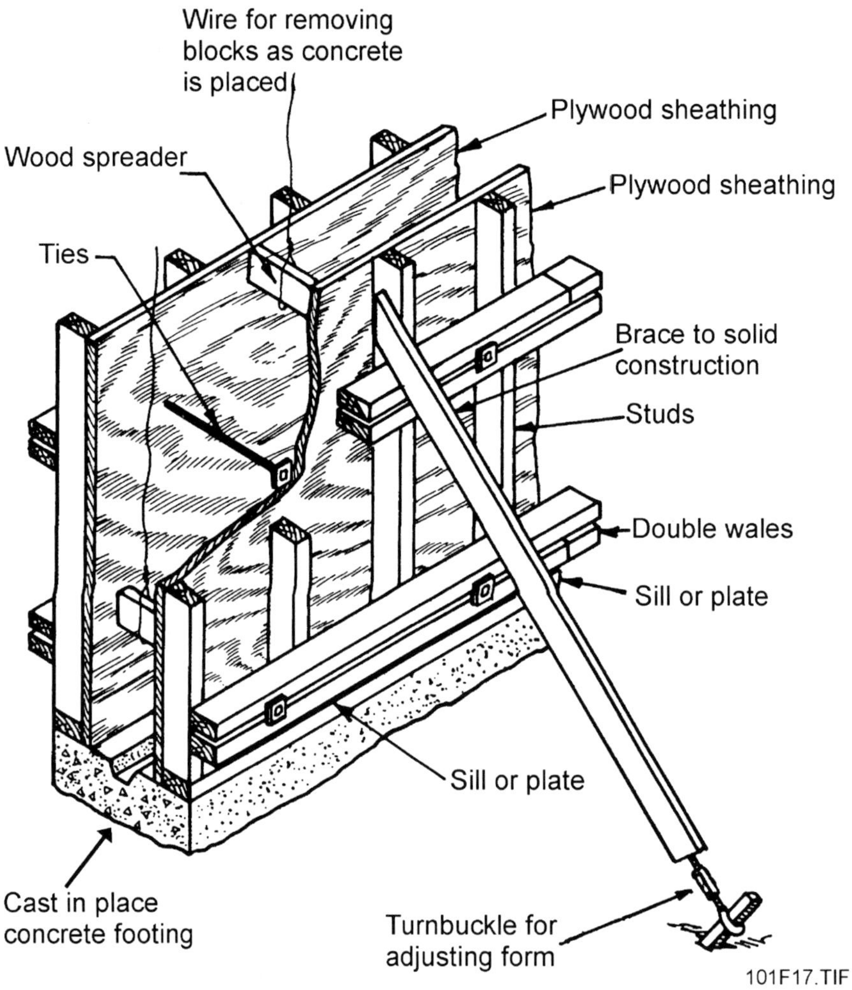

TRANSPARENCY 14

FIGURE 18

TASK MODULE 23101, INTRODUCTION TO CONCRETE CONSTRUCTION AND FINISHING

JOB-BUILT FORM FOR STEPS

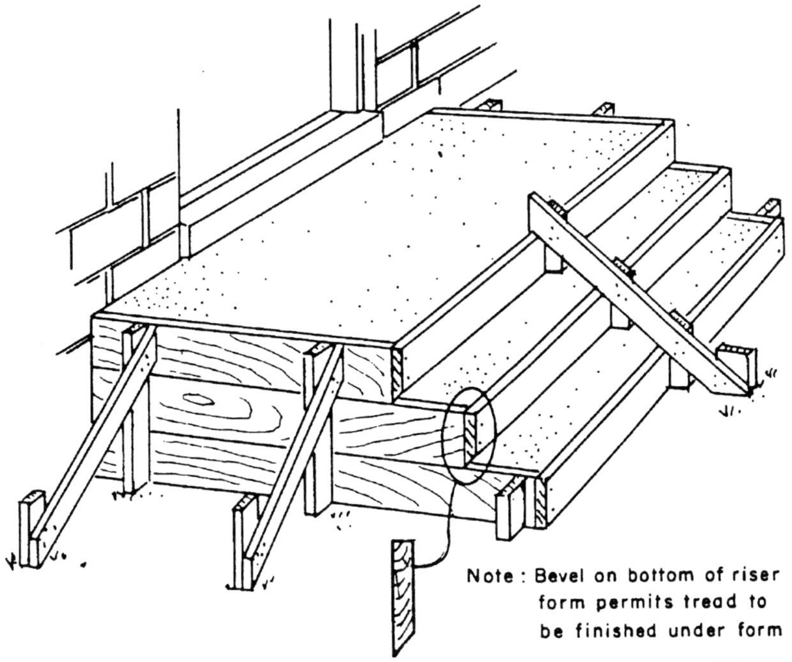

101F18.TIF

TRANSPARENCY 15

FIGURE 19

TASK MODULE 23101, INTRODUCTION TO CONCRETE CONSTRUCTION AND FINISHING

PREFABRICATED WALL FORM

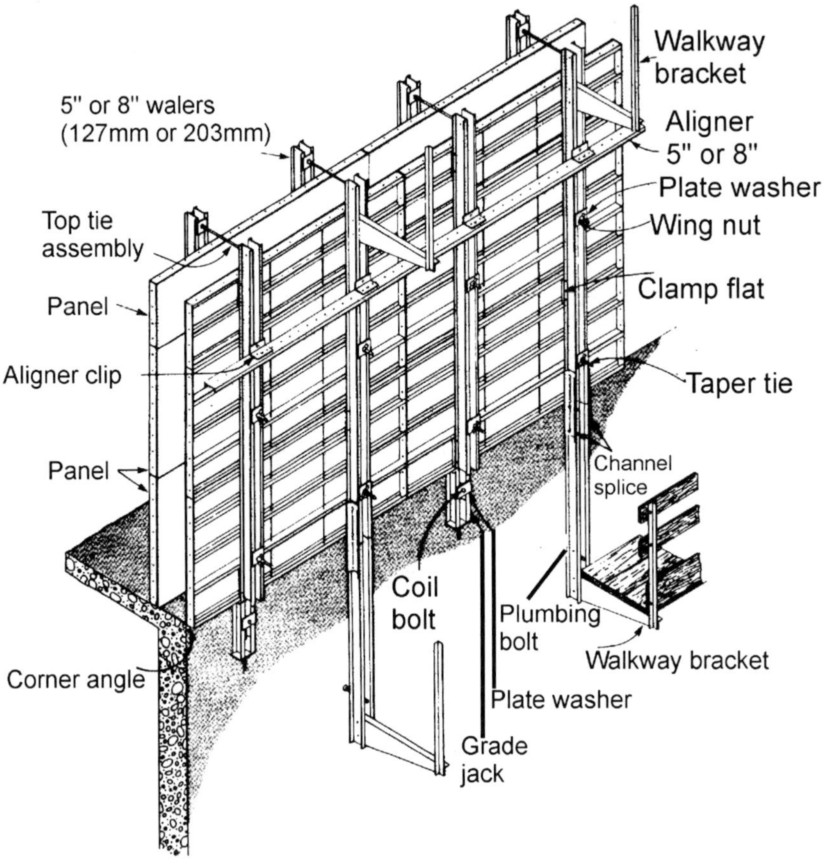

101F19.TIF

TRANSPARENCY 16

TASK MODULE 23101, INTRODUCTION TO CONCRETE CONSTRUCTION AND FINISHING

PLANT EXIT

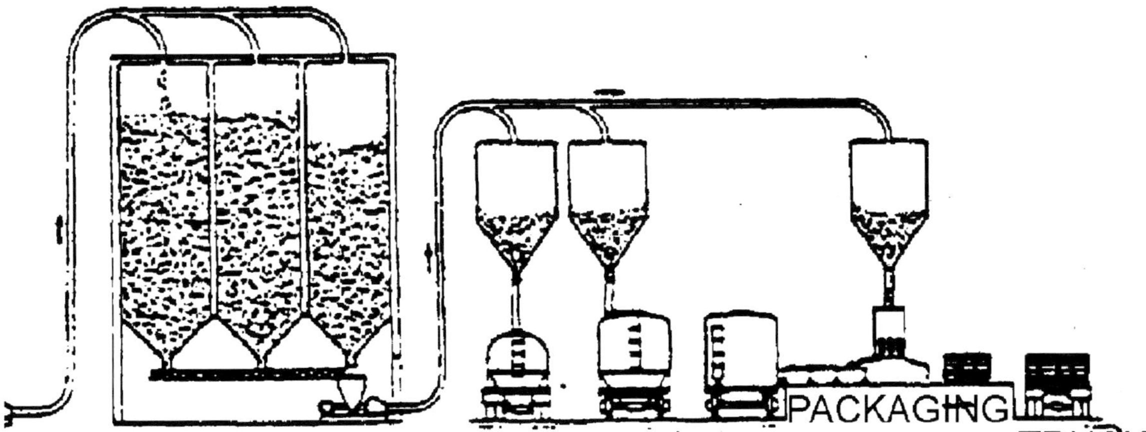

TR16IG.TIF

TRANSPARENCY 17

FIGURE 27

TASK MODULE 23101, INTRODUCTION TO CONCRETE CONSTRUCTION AND FINISHING

JOINTS IN A CONCRETE SLAB

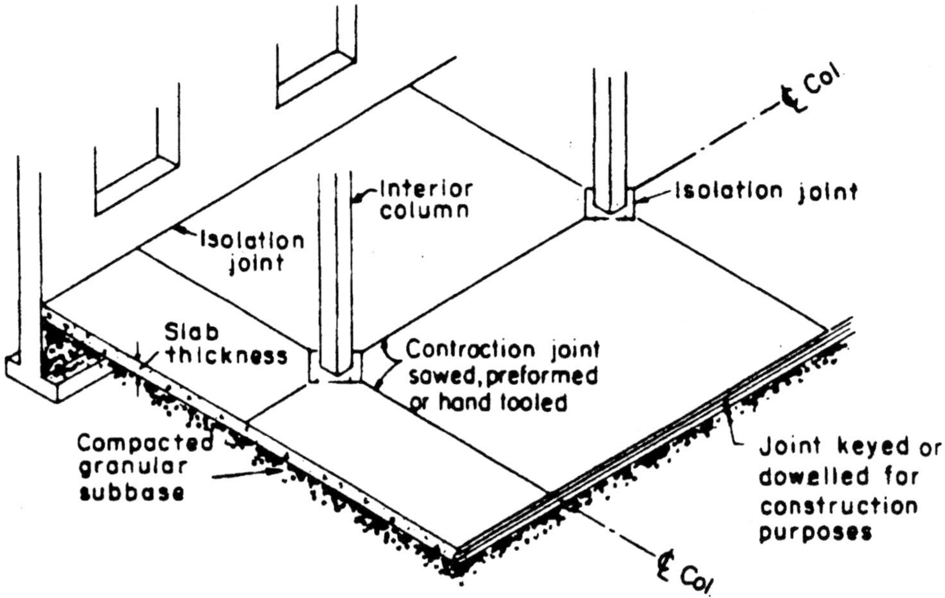

101F27.TIF

TRANSPARENCY 18

FIGURE 28

TASK MODULE 23101, INTRODUCTION TO CONCRETE CONSTRUCTION AND FINISHING

CURING WITH SPRAYED WATER TO KEEP CONCRETE MOIST

101F28.TIF

TRANSPARENCY 19

TASK MODULE 23101, INTRODUCTION TO CONCRETE CONSTRUCTION AND FINISHING

CRAFTWORKER KNOWLEDGE

Concrete craftworkers should know:

- What materials go into concrete, and how they react with each other
- How to safely assemble and take apart forms
- How to install reinforcing steel
- How to transport concrete by wheelbarrow, buggy, conveyor, bucket, or pump

TRANSPARENCY 20

TASK MODULE 23101, INTRODUCTION TO CONCRETE CONSTRUCTION AND FINISHING

CRAFTWORKER KNOWLEDGE (Cont.)

- How to use hand and power screeds, compactors, floats, trowels, jointers, edgers and finishing tools, and equipment on the job
- Curing methods and techniques
- The safety precautions necessary to protect themselves and other workers on the job

NOTE ON PERFORMANCE PROFILE TESTING

Performance Profiles are usually included in Instructor's Guides in a format that can be easily photocopied for each trainee. The Profiles measure trainee competency in the tasks taught in each module. There will be no Performance Profile Testing in this module.

NCCER CRAFT TRAINING USER UPDATES

The NCCER makes every effort to keep these manuals up-to-date and free of technical errors. We appreciate your help in this process. If you have an idea for improving this manual, or if you find an error, a typographical mistake, or an inaccuracy in the NCCER's Craft Training Manuals, please write us, using this form or a photocopy. Be sure to include the exact module number, page number, a description of the problem, and the correction, if possible. Your input will be brought to the attention of the Technical Review Committee. Thank you for your assistance.

Instructors – If you found that additional materials were necessary in order to teach this module effectively, please let us know so that we may include them in the Equipment/Materials list in the Instructor's Guide.

Write: Curriculum and Revision Department
National Center for Construction Education and Research
P.O. Box 141104
Gainesville, FL 32614-1104

Fax: 352-334-0932

Craft ______________________ Module Name ______________________

Module Number ______________________ Page Number(s) ______________________

Description of Problem

____

____

____

____

(Optional) Correction of Problem

____

____

____

(Optional) Your Name and Address

____

____

____

notes

Safety Requirements

Module 23102

SAFETY REQUIREMENTS

NATIONAL
CENTER FOR
CONSTRUCTION
EDUCATION AND
RESEARCH

OBJECTIVES

Upon completion of this module, the trainee will be able to:

1. Describe and wear different types of safety gear for the work site.
2. State the guidelines for dressing appropriately for concrete work.
3. Describe how to safely handle concrete when forming, placing, curing, and finishing.
4. Describe safety precautions to follow when working in extreme heat and cold.
5. Describe safety precautions to follow when working with hazardous materials.
6. Describe proper procedures for handling and maintaining concrete construction tools safely.

Note to the Instructor

Before teaching this module, you should review the details in this Instructor's Guide for Equipment and Supplies, Testing, and the suggested Teaching Sequence. Be sure to allow ample time to prepare your own training plan or lesson plan and to gather all required equipment and materials.

Required Equipment and Materials

The following are required for instruction using this module:

Equipment

Overhead projector and screen
Whiteboard/chalkboard
Appropriate Personal Protective Equipment:

Hard hats	Hearing protectors
Safety glasses	Face shields
Knee pads	Knee boards
Rubber gloves	Rubber boots
Safety shoes	Dust masks
Respirators	Safety harnesses

Concrete hand tools
Power float or trowel
Hand sprayer
Concrete saw
Carbon monoxide meter

Materials

Trainee Task Module
Pencils and paper
Markers/chalk
OSHA standards
Company safety manual
Operator's manuals for concrete equipment
Material Safety Data Sheets
Transparencies
Module Examinations
Performance Profile Sheets

HOW TO USE THIS INSTRUCTOR'S GUIDE

For each 2½ hour class session in this Instructor's Guide, the basic Presentation Sequence is as follows:

Introduction/Overview

Classroom, and/or Demonstration, and/or Laboratory

Class Break

Classroom, and/or Demonstration, and/or Laboratory

Summary

Suggested time periods for classroom sessions are included throughout this Instructor's Guide. These time periods should be adapted to meet local conditions and training requirements.

Each class session is presented with two columns of information. On the left side of the page, a narrow column provides suggested trainee and instructor actions, icons to call your attention to material, safety, audiovisual, or testing requirements, and space for your notes. The right-hand column provides the outline of the suggested presentation for each class session.

In this Instructor's Guide, the terms classroom, demonstration, and laboratory are defined and used as follows.

Classroom: Sessions are designed for lectures, group discussions, coaching, and additional activities. Trainees should be encouraged to actively participate.

Demonstration: Instructors will demonstrate all procedures before trainees attempt them. Instructors should make sure that trainees can point out all safety procedures during demonstrations to be assured of the proper use of equipment by trainees.

Laboratory: Instructors will facilitate all laboratory activities, coach trainees as they practice the procedures, monitor trainee progress, and provide feedback. The instructor will make sure that safety rules are followed at all times and that protective equipment is worn.

NCCER Standardized Craft Training Programs

The National Center for Construction Education and Research (NCCER) provides a standardized national program of accredited craft training. Key features of the program include instructor certification, competency-based training, and performance testing. The program provides trainees, instructors, and companies with a standard form of recognition through a National Craft Training Registry. The program is described in full in the *Guidelines for Accreditation*, published by the NCCER. For more information on standardized craft training, contact the NCCER at P.O. Box 141104, Gainesville, FL 32614-1104; or call 352-334-0911.

MODULE OVERVIEW

This module provides an overview of safety requirements specific to Concrete Finishing activities. It covers general work site safety, use and care of hand and power tools, hazardous and toxic chemicals, and the importance of appropriate personal protective equipment.

Prerequisites

Please see the Course Map. Prior to training with this module, it is recommended that the trainee shall have successfully completed the following modules:

Core Curricula

Concrete Finishing Level 1, Module 23101

Safety Considerations

Ensure that the trainees are equipped with Appropriate Personal Protective Equipment.

Teaching Time for This Module

Approximately 5 hours or two sessions of training time is suggested to cover *Safety Requirements*. The training class session is a suggested 2½ hour time period, which includes at least one break. **You will need to adjust the time required for hands-on activities and testing based on your class size and resources.** All time periods for this module are suggested and you will need to adapt the suggested lesson plan to meet your local conditions.

Suggested Teaching Sequence — Two 2½-Hour Sessions

Adjust your class times based on class size and resources.

Session	Topic	Trainee Module Section(s)
1	Introduction; Accident Prevention and First Aid Demonstration of Safety Equipment	1.0.0 – 4.4.0
2	Environmental Conditions; Surface Grinders Demonstration of Care of Tools Module Examination Performance Profile Examination	5.0.0 – 6.5.8

Optional References for Advanced Study

This module is intended to present thorough resources for task training. The following reference works are suggested for both instructors and motivated trainees interested in further study. These are optional materials for continued education rather than for task training.

Concrete Construction and Estimating, Craig Avery, Craftsman Book Company, Carlsbad, CA, 1991.

Concrete Manual, International Conference of Building Officials, Whittier, CA, 1992.

Concrete Parking Lots: Eight Steps to Quality Construction, Videotape, Portland Cement Association, Skokie, IL.

Concrete Technology, Third Edition, George R. White, Delmar Publishers, Inc., Albany, NY, 1991.

Design and Control of Concrete Mixtures, Thirteenth Edition, Portland Cement Association, Skokie, IL, 1988.

Finishing Concrete Flatwork, Videotape 32:00 minutes, Portland Cement Association, Skokie, IL, 1991.

Guide for Concrete Floor and Slab Construction, ACI 302.1R-96, American Concrete Institute, Farmington Hills, MI, 1997.

The Contractor's Guide to Quality Construction, American Concrete Institute, Farmington Hills, MI, 1997.

COURSE MAP

This course map shows all of the modules in the first level of the Concrete Finishing curricula. The suggested training order begins at the bottom and proceeds up. Skill levels increase as a trainee advances on the course map. The training order may be adjusted by the local Training Program Sponsor.

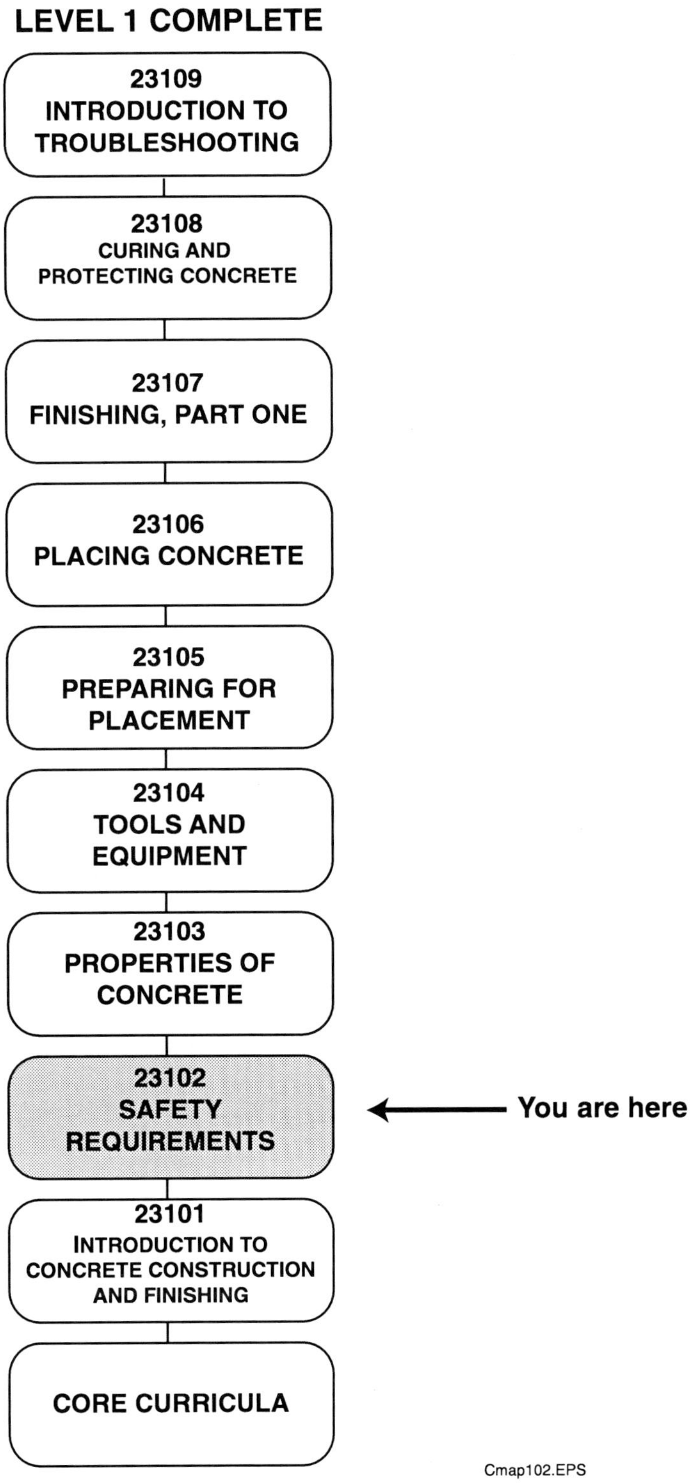

Cmap102.EPS

SESSION 1 OF 2 SESSIONS

Ensure that all trainees have brought their Trainee Modules to class.

Equipment and Materials

Trainee Task Module
Overhead projector and screen
Whiteboard/chalkboard
Markers/chalk
Pencils and paper
Transparencies
Hard hats
Safety glasses
Face shields
Hearing protectors
Rubber gloves
Rubber boots
Knee pads
Knee boards
Safety shoes
Dust masks
Safety harnesses
Respirators
Hand tools
Carbon monoxide meter

Time Schedule (150 minutes)

10 Overview
70 Classroom
15 Class Break
45 Demonstration
10 Summary

Overview (10 minutes)

Show Transparency 1, Course Objectives.

Show Transparency 2, Performance Profile Tasks.

1. Take attendance and perform other administrative tasks.
2. Outline briefly the topics to be covered. Explain to the trainees that they must pass the Module Examination with a grade of at least 70 percent. Also, explain that they must pass the Performance Profile Examination by performing each task to the level of acceptance criteria.
3. Provide a brief explanation of the acceptance criteria that will be used in the Performance Profile Examination.

Classroom (70 minutes)

List terms on the whiteboard.

1. Discuss the terms introduced in this module.
2. Discuss the overall issue of safety.
   - In the home
   - Driving
   - On the job

Pass around a copy of the company safety manual.

   Refer to the company safety manual relative to safety on the job.
3. Ask the trainees what they think is the main safety problem on their job. Ask trainees if any of them have been injured on the job. If so, have them explain what happened.

Show Transparency 3, Results of an Accident.

4. Emphasize what the results of an accident could be.
5. Describe the various types of Appropriate Personal Protective Equipment used in concrete construction.

Show Transparency 4, Hard Hat (Figure 1, Trainee Module).

6. Explain the use, wear, and adjustment of a hard hat. Demonstrate the proper way to wear a hard hat.
7. Explain the use, wear, and adjustment of safety glasses and face shields.
8. Explain the use, wear, and adjustment of hearing protectors.

Show Transparency 5, Respirators and Masks (Figure 4, Trainee Module).

9. Explain the use, wear, and adjustment of respirators and dust masks. Demonstrate the proper fitting of respirators and dust masks.

Show Transparency 6, Properly Dressed Concrete Worker (Figure 9, Trainee Module).

10. Explain the use and wear of gloves, knee pads, knee boards, safety shoes, and rubber boots.

Make sure carbon monoxide meter is working properly before the demonstration.

11. Describe the dangers of carbon monoxide in an enclosed area. Explain how to test for carbon monoxide. Demonstrate carbon monoxide meters that your company uses.

Show Transparency 7, Concreting Operations.

12. Discuss the overall safety considerations of concreting operations.
13. Discuss the hazards associated with mixing concrete.
    - Concrete on the skin and in the eyes
    - Keep hands and arms out of mobile mixer

AUDIO/VISUAL

14. Explain problems with transporting concrete.
    - Maneuvering ready-mix trucks
    - Use of wheelbarrows and Georgia buggies
    - Other transporting methods
15. Discuss hazards during placement.
    - Concrete on skin and in the eyes
    - Damage to forms
    - Hose under pressure during pumping
    - Accidental dumping of load
16. Describe the hazards associated with the finishing process.
    - Exposure to fresh concrete
    - Use of hand tools such as bullfloats and straightedges
17. Describe safety precautions that need to be followed during the curing process.
    - Wear Appropriate Personal Protective Equipment
    - Monitor pressure in sprayers
    - Keep sprayer in good working order
    - Know the treatment required for any chemicals being used
18. Summarize the hazards associated with erecting, using, and removing formwork.
    - Proper design of formwork
    - Distribution of loads
    - Proper bracing and tie off
    - Deflection of forms
    - Waiting period before form removal

Show Transparency 8, Actions That Promote Safety.

Write suggestions on the whiteboard.

19. Discuss ways that work site safety can be improved in your organization. Ask trainees for their suggestions.

Show Transparency 9, Positioning a Straight Ladder (Figure 11, Trainee Module).

20. Describe the proper procedure for setting up and using ladders.
21. Explain the four-point safety check which should be performed before using a ladder.

Show Transparency 10, Tubular Scaffold (Figure 12, Trainee Module).

22. Briefly describe the setting up and use of scaffolds. Explain the safety requirements for any scaffold.

23. Discuss the twelve unsafe actions and twelve unsafe conditions on a job site and what can be done to avoid them.
24. Discuss emergency procedures used in your company.
    - Giving first aid and CPR
    - Contacting the fire and rescue squad
    - Contacting police

Class Break (15 minutes)

Demonstration (45 minutes)

Arrange demonstration with vendor before session. Make sure vendor is prepared to demonstrate use of the safety harness.

1. Demonstrate the use of a safety harness. Have a safety consultant or vendor who sells safety equipment give a talk about the use and maintenance of harnesses and demonstrate how a harness is used.
2. Have trainees practice putting on and wearing safety harnesses.

Check all respirators before beginning the demonstration.

3. Demonstrate proper fitting of respirators.
4. Have trainees practice putting on and wearing respirators.

Summary (10 minutes)

1. Review the key concepts of this session.
2. Answer trainee questions.
3. Give assignments for the next session.

SESSION 2 OF 2 SESSIONS

Ensure that all trainees have brought their Trainee Modules to class.

Equipment and Materials

Trainee Task Module
Overhead projector and screen
Whiteboard/chalkboard
Markers/chalk
Pencils and paper
Material Safety Data Sheet
Transparencies
Appropriate Personal Protective Equipment
Concrete hand tools
Straightedge
Float
Trowel
Power float or trowel
Safety harnesses
Respirators
Module Examinations
Performance Profile Sheets

Time Schedule (150 minutes)

10 Overview
50 Classroom/Demonstration
5 Class Break
25 Classroom
5 Class Break
30 Module Examination
5 Class Break
20 Performance Profile Examination

Overview (10 minutes)

Show Transparency 1, Course Objectives.

Show Transparency 2, Performance Profile Tasks.

1. Take attendance and perform other administrative tasks.
2. Outline briefly the topics to be covered. Explain to the trainees that they must pass the Module Examination with a grade of at least 70 percent. Also, explain that they must pass the Performance Profile Examination by performing each task to the level of acceptance criteria.
3. Provide a brief explanation of the acceptance criteria that will be used in the Performance Profile Examination.

Classroom/Demonstration (50 minutes)

1. Discuss the problems with working under extreme weather conditions.

Show Transparency 11, Wind Chill Chart (Figure 15, Trainee Module).

2. Discuss the effects that cold weather has on concrete and on the concrete workers. Explain the phenomenon of wind chill and how it can affect your body temperature.
3. Discuss the dangers of working in heated enclosures.
4. Discuss the effects that extreme hot weather has on concrete and the concrete worker.
5. Describe what precautions to take during hot weather.
6. Describe precautions that should be taken during other environmental conditions.
   - High winds and rain
   - Thunderstorms
   - Snow
7. Explain any policies your company has regarding work during extreme weather conditions.

Pass around an example of a Material Safety Data Sheet.

8. Discuss the requirements for hazardous material labeling and OSHA standards for Hazard Communication.
9. Discuss the handling of toxic substances.

Use hand tools to demonstrate hazards.

10. Discuss the potential hazards of various hand tools used in concrete construction. Describe precautions that should be taken.
    - Straightedges
    - Floats
    - Trowels

Use power float or trowel to demonstrate hazard points.

11. Discuss the potential hazards of various power tools used in concrete construction. Describe precautions.

Class Break (5 minutes)

Classroom (25 minutes)

1. Have trainees complete the Review/Practice Questions found in the back of the Trainee Task Module.
2. Review the answers and address any comments or questions the trainees may have.
3. Tell trainees to prepare for taking the Module Examination and Performance Profile Examination.

Class Break (5 minutes)

TESTING

Module Examination (30 minutes)

1. Hand out Module Examinations to all trainees and tell them they have 30 minutes to complete the exam.
2. Collect and grade examinations.

Class Break (5 minutes)

Performance Profile Examination (20 minutes)

1. Have trainees perform the following performance objectives to the satisfaction of the instructor:
   - Lay out, inspect, and put on a safety harness with necessary strap adjustments.
   - Check operation of respirators, place on face and adjust properly.
2. Fill out Performance Profile Sheets for each trainee.

MODULE EXAMINATIONS AND ANSWER KEYS

The masters for Module Examinations and Answer Keys are found in the booklet entitled "Concrete Finishing Level One Module Examinations" included in this Instructor's Guide package. You may also reproduce the tests using the Windows-compatible software program which may be purchased separately from Prentice Hall. If you do not have the reproducible masters in your Instructor's Guide package, or would like to order the test software, please contact Prentice Hall Publishers at 1-800-922-0579.

Be sure to keep answer keys and tests in a safe area.

OVERHEAD TRANSPARENCIES

Masters for overhead transparencies are found on the following pages. Transparencies can be made by using clear acetate sheets in a photocopy machine. This can be done at any commercial copy shop, or by following this method on your office machine:

1. Use standard $8^{1}/_{2} \times 11$ clear acetate sheets, which can be obtained at office supply stores, stationery stores, or commercial copying centers.
2. Check the photocopier's manual to make sure it will take the thicker acetate sheets.
3. Remove the paper supply in the photocopying machine and replace it with a supply of acetate sheets.
4. Make photocopies from the masters on the following pages in the usual way.
   The material will be reproduced on the clear acetate sheets for use on any overhead projector.

TRANSPARENCY 1

TASK MODULE 23102, SAFETY REQUIREMENTS

COURSE OBJECTIVES

Upon completion of this module, the trainee will be able to:

1. **Describe and wear different types of safety gear for the work site.**
2. **State the guidelines for dressing appropriately for concrete work.**
3. **Describe how to safely handle concrete when forming, placing, curing, and finishing.**
4. **Describe safety precautions to follow when working in extreme heat and cold.**
5. **Describe safety precautions to follow when working with hazardous materials.**
6. **Describe proper procedures for handling and maintaining concrete construction tools safely.**

TRANSPARENCY 2

TASK MODULE 23102, SAFETY REQUIREMENTS

PERFORMANCE PROFILE TASKS

1. Layout, inspect, and put on a safety harness with necessary strap adjustments.

2. Check operation of respirators, place on face, and adjust properly.

TRANSPARENCY 3

TASK MODULE 23102, SAFETY REQUIREMENTS

RESULTS OF AN ACCIDENT

- Injury with lost time from work.
- Loss of hearing, sight, or limb.
- Medical bills.
- Loss of financial stability.
- Loss of life.

TRANSPARENCY 4

FIGURE 1

TASK MODULE 23102, SAFETY REQUIREMENTS

HARD HAT

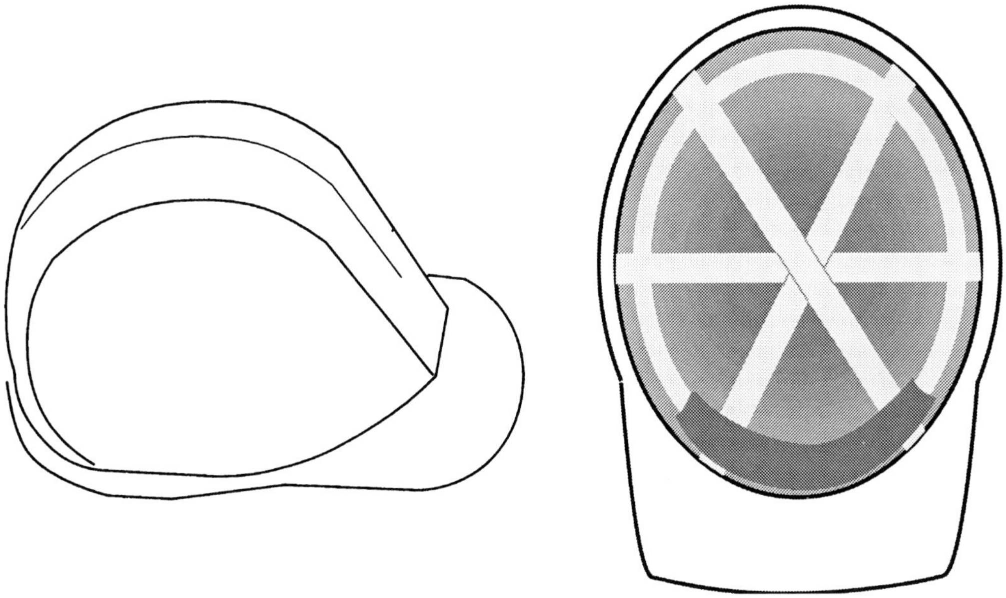

102F01.tif

TRANSPARENCY 5

FIGURE 4

TASK MODULE 23102, SAFETY REQUIREMENTS

RESPIRATORS AND MASKS

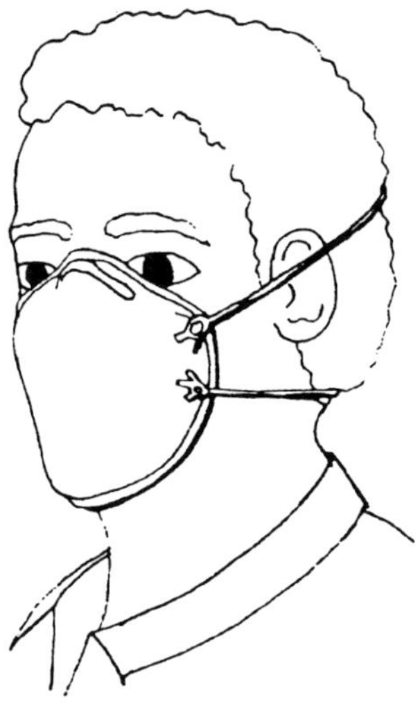

102F04.tif

TRANSPARENCY 6

FIGURE 9

TASK MODULE 23102, SAFETY REQUIREMENTS

PROPERLY DRESSED CONCRETE WORKER

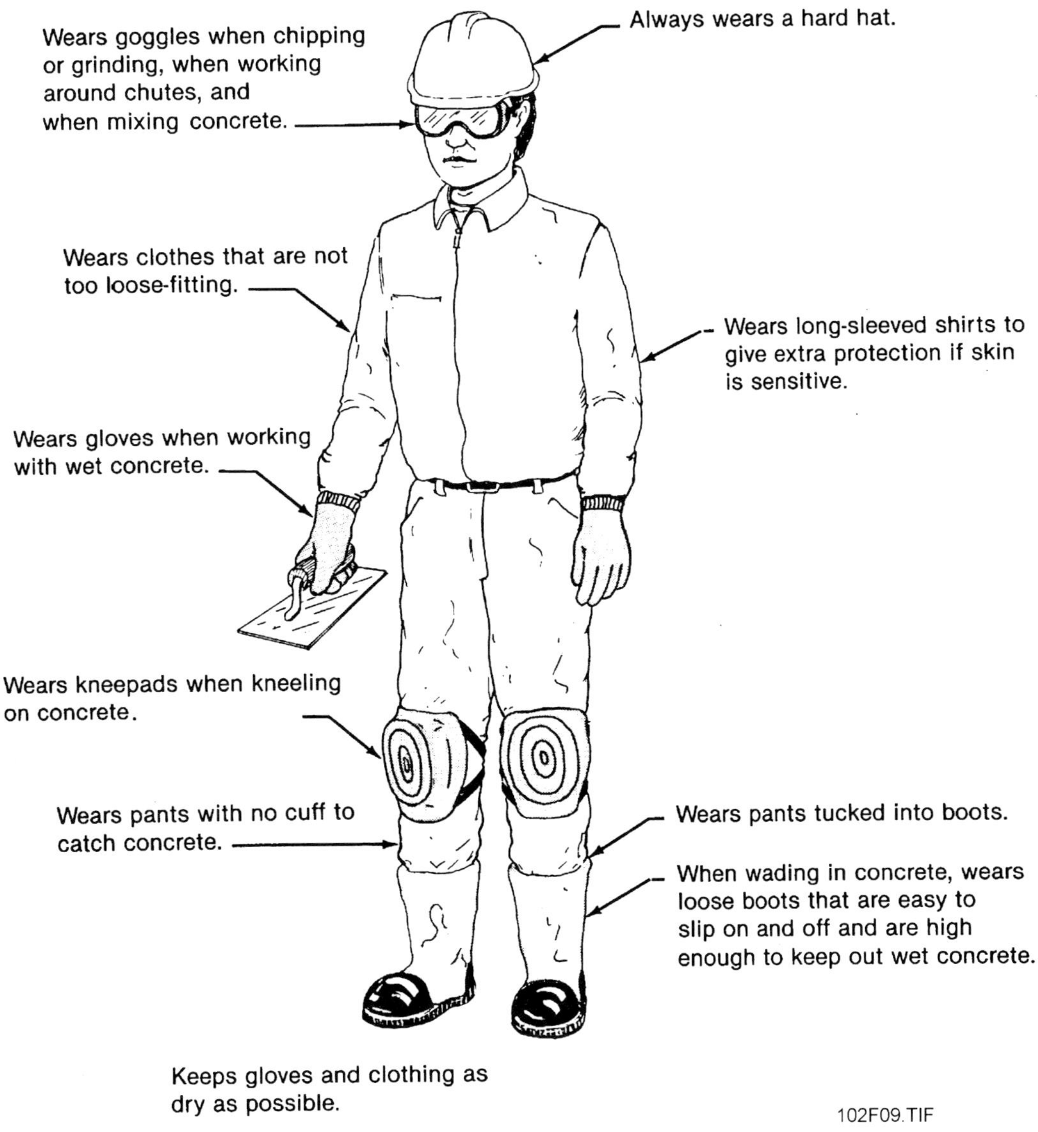

TRANSPARENCY 7

TASK MODULE 23102, SAFETY REQUIREMENTS

CONCRETING OPERATIONS

- Mixing
- Transporting
- Placing
- Finishing
- Curing
- Formwork

TRANSPARENCY 8

TASK MODULE 23102, SAFETY REQUIREMENTS

ACTIONS THAT PROMOTE SAFETY

- Maintain job site in clean and orderly condition.
- Encourage and enforce personal hygiene.
- Keep all work surfaces dry and slip-resistant.
- Store combustible scrap and waste safely and dispose promptly.
- Remove accumulations of dirt.

TRANSPARENCY 9

FIGURE 11

TASK MODULE 23102, SAFETY REQUIREMENTS

POSITIONING A STRAIGHT LADDER

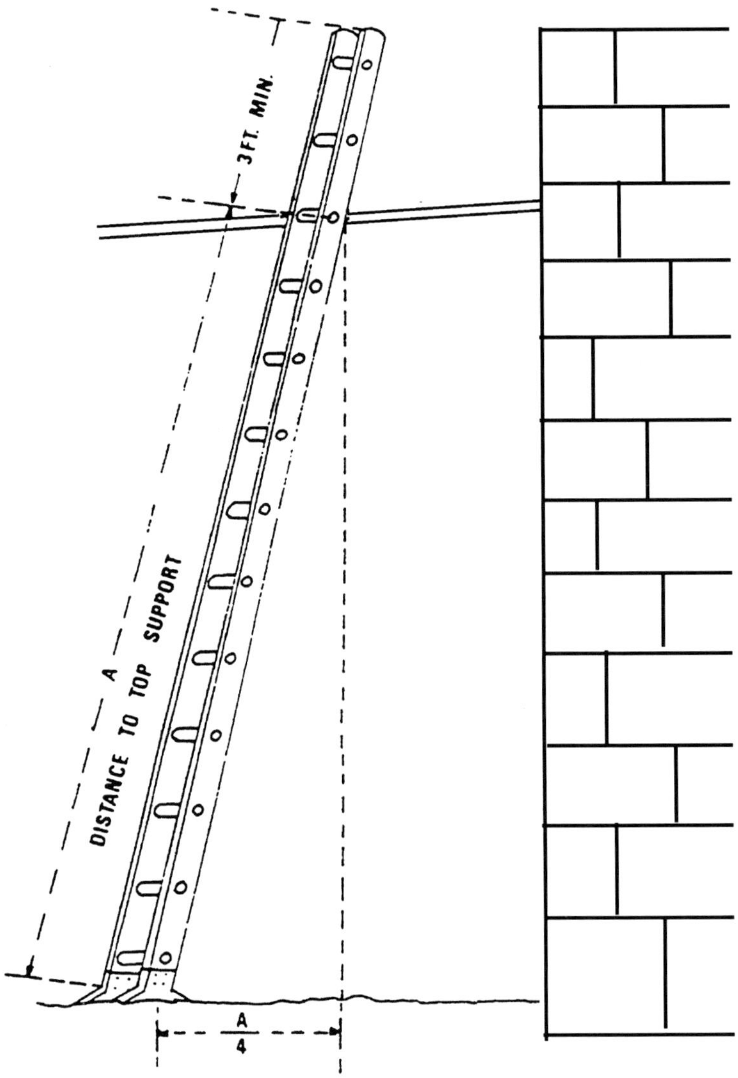

102F11.TIF

TRANSPARENCY 10

FIGURE 12

TASK MODULE 23102, SAFETY REQUIREMENTS

TUBULAR SCAFFOLD

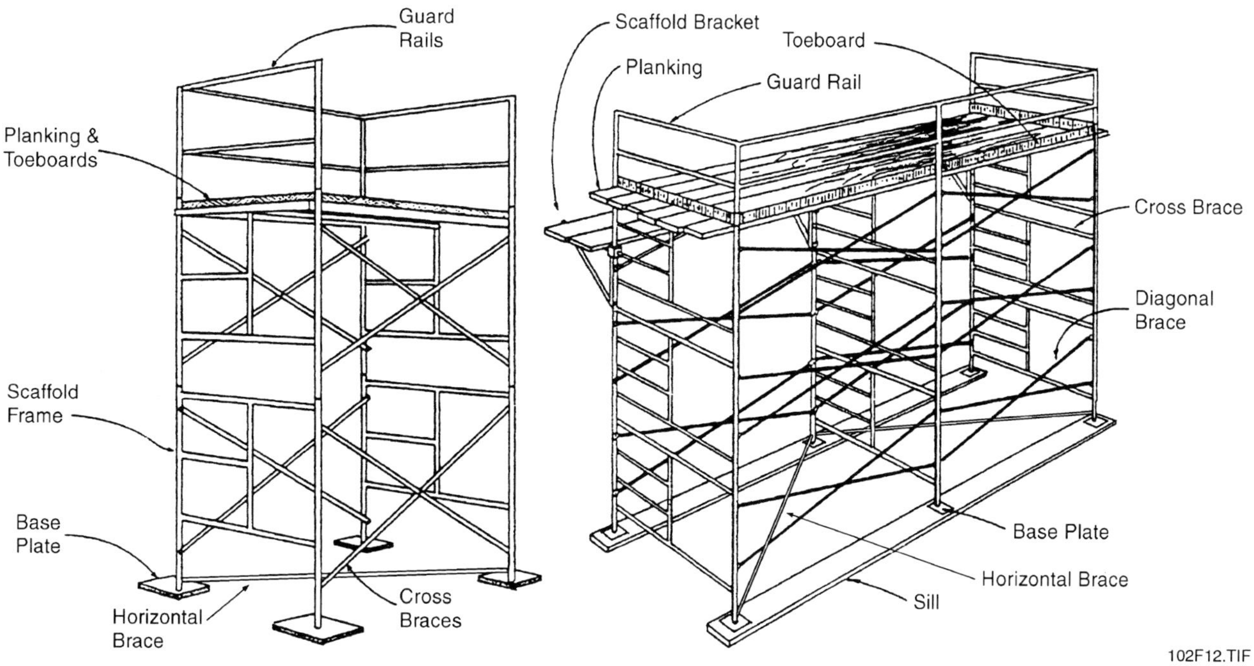

102F12.TIF

TRANSPARENCY 11

FIGURE 15

TASK MODULE 23102, SAFETY REQUIREMENTS

WINDCHILL CHART

WIND-CHILL CHART							
Estimated Wind Speed MPH	ACTUAL THERMOMETER READING (DEGREES F)						
	40	30	20	10	0	-10	-20
10	28	16	4	-9	-21	-33	-46
20	18	4	-10	-25	-39	-53	-67
30	13	-2	-18	-33	-48	-63	-79
40	10	-6	-21	-37	-53	-69	-85
Wind speeds greater than 40 MPH have little additional effect	Little danger for properly clothed person			Increasing danger			Great danger
				Danger from freezing of exposed flesh			

102F15.EPS

NOTE ON PERFORMANCE PROFILE TESTING

Performance Profiles are included in this Instructor's Guide in a format that can be easily photocopied for each trainee. The Profiles measure trainee competency in the tasks taught in this module.

Please note the number of tasks to be tested while teaching this module. Each trainee should be tested on the tasks listed in the Performance Profile. Before the performance testing, the instructor should brief the trainees on:

- test objectives and criteria,
- safety precautions, and
- procedures for each task to be tested.

The instructor administering the performance testing should also do the following:

- ensure that all of the needed equipment is available and operating properly;
- set up the testing stations;
- organize and administer the test in a way that allows for optimal performance;
- complete the Performance Profile Sheet for each trainee by assigning a score for each listed task;
- monitor adherence to all safety regulations and precautions;
- provide adequate supervision to prevent injuries; and
- take immediate and effective action to remedy any emergency.

ACCREDITATION TESTING

If this Performance Profile Testing is done in the National Center for Construction Education and Research Standardized Craft Training Program, the following conditions must be met:

1. The Craft Instructor must hold valid NCCER instructor certification.
2. The training must be delivered through a Training Program Sponsor recognized by the NCCER.
3. For every module, the specific performance testing must be completed to the satisfaction of the instructor.
4. The results of the testing must be recorded on the Craft Training Report Form. This form must be provided to the local Training Program Sponsor to be forwarded to the NCCER National Craft Training Registry.

Craft: Concrete Finishing	**WHEELS**
Task Module Number: 23102	**OF**
Task Module Title: Safety Requirements	**LEARNING**

TRAINEE NAME: ______________________________

TRAINEE SOCIAL SECURITY NUMBER: ______________________________

CLASS: ______________________________

TRAINING PROGRAM SPONSOR: ______________________________

____

INSTRUCTOR: ______________________________

Rating Levels: 1. Passed: performed task.

2. Failed: did not perform task.

Recognition: When testing for the NCCER Standardized Craft Training Program, be sure to record Performance Profile testing results on Craft Training Report Form 200 and submit the results to the Training Program Sponsor.

Objective	TASK	RATING
1	1. Layout, inspect, and put on a safety harness with necessary strap adjustments.	
1	2. Check operation of respirators, place on face, and adjust properly.	

NCCER CRAFT TRAINING USER UPDATES

The NCCER makes every effort to keep these manuals up-to-date and free of technical errors. We appreciate your help in this process. If you have an idea for improving this manual, or if you find an error, a typographical mistake, or an inaccuracy in the NCCER's Craft Training Manuals, please write us, using this form or a photocopy. Be sure to include the exact module number, page number, a description of the problem, and the correction, if possible. Your input will be brought to the attention of the Technical Review Committee. Thank you for your assistance.

Instructors – If you found that additional materials were necessary in order to teach this module effectively, please let us know so that we may include them in the Equipment/Materials list in the Instructor's Guide.

Write: Curriculum and Revision Department
National Center for Construction Education and Research
P.O. Box 141104
Gainesville, FL 32614-1104

Fax: 352-334-0932

Craft ______________________ Module Name ______________________

Module Number ______________________ Page Number(s) ______________________

Description of Problem

____

____

____

____

(Optional) Correction of Problem

____

____

____

(Optional) Your Name and Address

____

____

____

notes

Properties of Concrete

Module 23103

PROPERTIES OF CONCRETE

NATIONAL
CENTER FOR
CONSTRUCTION
EDUCATION AND
RESEARCH

OBJECTIVES

Upon completion of this module, the trainee will be able to:

1. Describe the properties of concrete.
2. Explain how the properties of concrete are used in construction.
3. Determine how the ingredients of concrete influence mix, placement, finishing, durability, and performance.
4. Describe quality-control tests on concrete ingredients, fresh concrete, and hardened concrete.
5. Mix a test batch of concrete.
6. Perform a slump test.

Note to the Instructor

Before teaching this module, you should review the details in this Instructor's Guide for Equipment and Supplies, Testing, and the suggested Teaching Sequence. Be sure to allow ample time to prepare your own training plan or lesson plan and to gather all required equipment and materials.

Required Equipment and Materials

The following are required for instruction using this module:

Equipment
Overhead projector and screen
Whiteboard/chalkboard
Mixing boards (plywood)
Hoes
Square-end shovels
Buckets or cubic foot boxes
Slump test kits
Appropriate Personal Protective Equipment

Materials
Trainee Task Module
Transparencies
Pencils and paper
Markers/chalk
Example specification sheets
Example reinforcing steel bars
Portland cement
Sand
Aggregate
Water
Admixture samples
Module Examinations
Performance Profile Sheets

HOW TO USE THIS INSTRUCTOR'S GUIDE

For each 2½ hour class session in this Instructor's Guide, the basic Presentation Sequence is as follows:

Introduction/Overview

Classroom, and/or Demonstration, and/or Laboratory

Class Break

Classroom, and/or Demonstration, and/or Laboratory

Summary

Suggested time periods for classroom sessions are included throughout this Instructor's Guide. These time periods should be adapted to meet local conditions and training requirements.

Each class session is presented with two columns of information. On the left side of the page, a narrow column provides suggested trainee and instructor actions, icons to call your attention to material, safety, audiovisual, or testing requirements, and space for your notes. The right-hand column provides the outline of the suggested presentation for each class session.

In this Instructor's Guide, the terms classroom, demonstration, and laboratory are defined and used as follows.

Classroom: Sessions are designed for lectures, group discussions, coaching, and additional activities. Trainees should be encouraged to actively participate.

Demonstration: Instructors will demonstrate all procedures before trainees attempt them. Instructors should make sure that trainees can point out all safety procedures during demonstrations to be assured of the proper use of equipment by trainees.

Laboratory: Instructors will facilitate all laboratory activities, coach trainees as they practice the procedures, monitor trainee progress, and provide feedback. The instructor will make sure that safety rules are followed at all times and that protective equipment is worn.

NCCER Standardized Craft Training Programs

The National Center for Construction Education and Research (NCCER) provides a standardized national program of accredited craft training. Key features of the program include instructor certification, competency-based training, and performance testing. The program provides trainees, instructors, and companies with a standard form of recognition through a National Craft Training Registry. The program is described in full in the *Guidelines for Accreditation*, published by the NCCER. For more information on standardized craft training, contact the NCCER at P.O. Box 141104, Gainesville, FL 32614-1104; or call 352-334-0911.

MODULE OVERVIEW

This module introduces the trainee to the properties of concrete and the components that make up the concrete mixture. Chemical and physical properties of cement, aggregate, and various types of admixtures are described. Basic tests used to determine properties such as slump and ultimate strength are explained.

Prerequisites

Please see the Course Map. Prior to training with this module, it is recommended that the trainee shall have successfully completed the following modules:

Core Curricula

Concrete Finishing Level 1, Modules 23101 and 23102

Safety Considerations

Ensure that the trainees are equipped with Appropriate Personal Protective Equipment.

Teaching Time for This Module

Approximately 10 hours or four sessions of training time are suggested to cover *Properties of Concrete*. The training class session is a suggested 2½ hour time period, which includes at least one break. **You will need to adjust the time required for hands-on activities and testing based on your class size and resources.** All time periods for this module are suggested, and you will need to adapt the suggested lesson plan to meet your local conditions.

Suggested Teaching Sequence — Four 2½-Hour Sessions

Adjust your class times based on class size and resources.

Session	Topic	Trainee Module Section(s)
1	Introduction; Components; Concrete Production	1.0.0 – 3.3.3
2	Properties of Concrete; Testing and Quality Control Demonstration/Laboratory	4.0.0 – 5.3.4
3	Laboratory/Site Visit	2.0.0 – 5.2.0
4	Module Examination Performance Profile Examination	

Optional References for Advanced Study

This module is intended to present thorough resources for task training. The following reference works are suggested for both instructors and motivated trainees interested in further study. These are optional materials for continued education rather than for task training.

Concrete Construction and Estimating, Craig Avery, Craftsman Book Company, Carlsbad, CA, 1991.

Concrete Manual, International Conference of Building Officials, Whittier, CA, 1992.

Concrete Parking Lots: Eight Steps to Quality Construction, Videotape, Portland Cement Association, Skokie, IL.

Concrete Technology, Third Edition, George R. White, Delmar Publishers, Inc., Albany, NY, 1991.

Design and Control of Concrete Mixtures, Thirteenth Edition, Portland Cement Association, Skokie, IL, 1988.

Finishing Concrete Flatwork, Videotape 32:00 minutes, Portland Cement Association, Skokie, IL, 1991.

Guide for Concrete Floor and Slab Construction, ACI 302.1R-96, American Concrete Institute, Farmington Hills, MI, 1997.

The Contractor's Guide to Quality Construction, American Concrete Institute, Farmington Hills, MI, 1997.

COURSE MAP

This course map shows all of the modules in the first level of the Concrete Finishing curricula. The suggested training order begins at the bottom and proceeds up. Skill levels increase as a trainee advances on the course map. The training order may be adjusted by the local Training Program Sponsor.

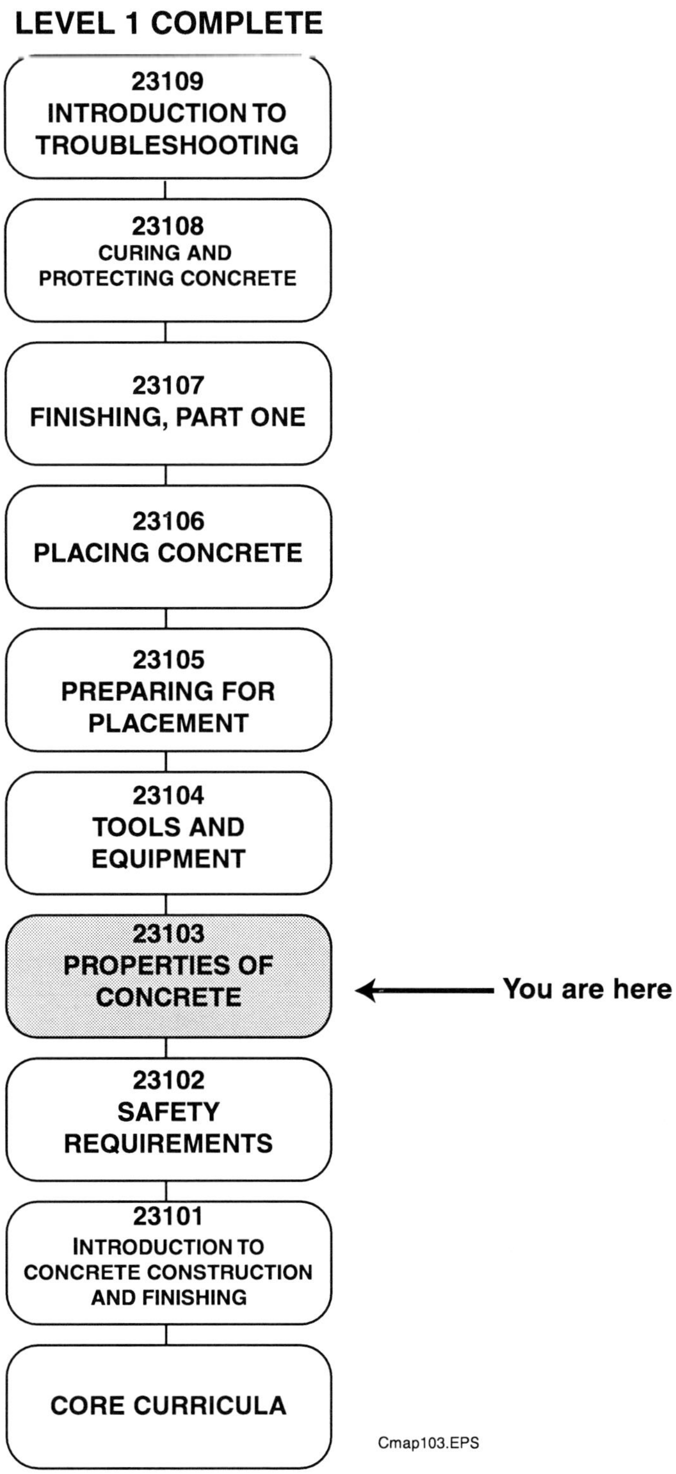

Cmap103.EPS

SESSION 1 OF 4 SESSIONS

Ensure that all trainees have brought their Trainee Modules to class.

Equipment and Materials

Trainee Task Module
Overhead projector and screen
Whiteboard/chalkboard
Markers/chalk
Transparencies
Pencils and paper
Admixture samples

Time Schedule (150 minutes)

15 Overview
55 Classroom
15 Class Break
55 Classroom
10 Summary

Show Transparency 1, Course Objectives.

Show Transparency 2, Performance Profile Tasks.

Overview (15 minutes)

1. Take attendance and perform other administrative tasks.
2. Outline briefly the topics to be covered. Explain to the trainees that they must pass the Module Examination with a grade of at least 70 percent. Also, explain that they must pass the Performance Profile Examination by performing each task to the level of acceptance criteria.
3. Provide a brief explanation of the acceptance criteria that will be used in the Performance Profile Examination.

Classroom (55 minutes)

Use the whiteboard to explain terms.

1. Discuss the trade terms introduced in this module and explain them.
2. Explain why it is important to understand the various properties of concrete.

Show Transparency 3, A Cubic Foot of Typical Concrete (Figure 1, Trainee Module).

3. Describe the various components of concrete.
   - Cement
   - Aggregate
   - Water
   - Admixtures

Show Transparency 4, Characteristics and Tests of Aggregates (Figure 2, Trainee Module).

4. Describe the characteristics of coarse aggregates.
5. Describe the characteristics of fine aggregates.
6. Discuss the tests used on aggregates.

AUDIO/VISUAL

Show Transparency 5, Cement Types.

7. Discuss the different types of portland cement and blended cements.
8. Discuss the characteristics of water needed for concrete mixing.

Class Break (15 minutes)

Classroom (55 minutes)

Show Transparencies 6 and 7, Concrete Admixtures By Classification.

1. Discuss the effects of admixtures on plastic and hardened concrete.
2. Discuss the properties and effects of chemical admixtures.

Show Transparency 8, Fly Ash Production (Figure 4, Trainee Module).

3. Discuss the properties and effects of mineral admixtures.

Show Transparencies 9 and 10, Effects of Entrained Air on Concrete Properties (Figure 5, Trainee Module).

4. Describe the effects of air-entraining admixtures.
5. Explain batching procedures.
   - Automatic
   - Semi-automatic
   - Manual
6. Describe mixing concrete by truck-mounted mixer and mobile batch mixer.

Show Transparency 11, Concrete Proportions by Volume (Figure 10, Trainee Module).

7. Discuss mixing concrete by hand. Explain how to properly proportion the mix.
8. Discuss mixing concrete using a small portable mixer.

Summary (10 minutes)

1. Review the materials covered in this session.
2. Answer trainee questions
3. Give assignment for the next session.

SESSION 2 OF 4 SESSIONS

Ensure that all trainees have brought their Trainee Modules to class.

Equipment and Materials

Trainee Task Module
Overhead projector and screen
Whiteboard/chalkboard
Markers/chalk
Transparencies
Pencils and paper
Mixing boards (plywood)
Square-end shovels
Buckets or cubic foot boxes
Slump test kits
Example reinforcing steel bars
Example specification sheets
Portland cement
Sand
Aggregate
Water
Appropriate Personal Protective Equipment

Time Schedule (150 minutes)

5 Overview
45 Classroom
10 Class Break
30 Demonstration
5 Class Break
45 Laboratory
10 Summary

Show Transparency 1, Course Objectives.

Show Transparency 2, Performance Profile Tasks.

Overview (5 minutes)

1. Take attendance and perform other administrative tasks.
2. Outline briefly the topics to be covered. Explain to the trainees that they must pass the Module Examination with a grade of at least 70 percent. Also, explain that they must pass the Performance Profile Examination by performing each task to the level of acceptance criteria.
3. Provide a brief explanation of the acceptance criteria that will be used in the Performance Profile Examination.

AUDIO/VISUAL

Classroom (45 minutes)

Show Transparency 12, Types of Strength (Figure 13, Trainee Module).

1. Discuss the types of strength required for concrete elements.
   - Compression
   - Tension
   - Flexure

Show Transparency 13, Concrete Age-Strength Relationships (Figure 14, Trainee Module).

2. Explain concrete age to strength relationship. Point out the difference between air-entrained and non-air-entrained concrete.

Show Transparency 14, Various Types Of Reinforcing Bars (Figure 16, Trainee Module).

3. Discuss reinforcing steel bars and their uses.
4. Discuss setting time and hydration of concrete.

Pass around examples of specification sheets from ACI or other organizations.

5. Explain the requirements for concrete specifications and how they are developed.
6. Discuss testing procedures and quality-control requirements as they relate to the properties of concrete materials.
   - Air content
   - Strength
   - Weight
   - Temperature

Show Transparency 15, Equipment for Slump Test (Figure 21, Trainee Module).

7. Explain the purpose of the slump test and explain how slump is measured.

Show Transparency 16, Slump Measurement (Figure 23, Trainee Module).

8. Review the steps in performing a slump test.
9. Describe the types of inspections that would be made of different concrete materials and components.
   - Forms
   - Reinforcement
   - Batching and mixing
   - On-site inspections

Class Break (10 minutes)

Make sure all materials and equipment are available before beginning the demonstration.

Demonstration (30 minutes)

1. Demonstrate mixing concrete manually. Use a basic mix design for your area.
2. Demonstrate performing a slump test on the concrete that you have mixed.

Class Break (5 minutes)

Laboratory (45 minutes)

1. Divide trainees into groups of two to four.
2. Have trainees practice mixing concrete by hand. Observe and monitor performance and make any necessary corrections.
3. Observe and monitor performance as trainees perform a slump test.
4. Ensure that all Appropriate Personal Protective Equipment is used correctly.

Summary (10 minutes)

1. Review the materials covered in this session.
2. Answer trainee questions.
3. Give assignments for the next session. Ask trainees to begin working on the review/practice questions in the Trainee Module.
4. Inform the trainees that the next session will include a field trip. Remind them to bring notebooks and writing materials so they can take notes of their visit.

SESSION 3 OF 4 SESSIONS

Equipment and Materials

Trainee Task Module
Pencils and paper
Appropriate Personal Protective Equipment

Ensure that all trainees have brought their Trainee Modules and writing materials to class.

Time Schedule (150 minutes)

15 Overview
120 Site Visit
15 Summary

Overview (15 minutes)

Arrange for the site visit well in advance. Make sure sufficient transportation is available for all trainees.

1. Take attendance and perform other administrative tasks.
2. Discuss taking notes and assign specific responsibilities.
3. Impress on trainees the importance of:
   - Staying alert
   - Staying with the group
   - Obeying instructions

Site Visit (120 minutes)

1. Tour a batch plant, pointing out items discussed in class.
2. Observe sampling and testing of materials.
3. Answer trainee questions.
4. Tour a ready-mix plant, pointing out items discussed in class.
5. Continue observation of sampling and testing of materials that may be different from the batch plant.
6. Answer trainee questions.

Summary (15 minutes)

1. Review the events and items observed.
2. Give assignments for the next session.
3. Explain that the next session will be the last session and will includes the Module Examination and the Performance Profile Examination.

SESSION 4 OF 4 SESSIONS

Ensure that all trainees have brought their Trainee Modules to class.

Equipment and Materials

Trainee Task Module
Overhead projector and screen
Whiteboard/chalkboard
Markers/chalk
Pencils and paper
Transparencies
Mixing boards (plywood)
Hoes
Square-end shovels
Buckets or cubic foot boxes
Slump test kits
Portland cement
Sand
Aggregate
Water
Appropriate Personal Protective Equipment
Module Examinations
Performance Profile Sheets

Time Schedule (150 minutes)

15 Overview
45 Laboratory
10 Class Break
30 Module Examination
5 Class Break
45 Performance Profile Examination

Overview (15 minutes)

Show Transparency 1, Course Objectives.

Show Transparency 2, Performance Profile Tasks.

1. Take attendance and perform other administrative tasks.
2. Outline briefly the topics to be covered. Explain to the trainees that they must pass the Module Examination with a grade of at least 70 percent. Also, explain that they must pass the Performance Profile Examination by performing each task to the level of acceptance criteria.
3. Provide a brief explanation of the acceptance criteria that will be used in the Performance Profile Examination.

Laboratory (45 minutes)

1. Observe and monitor trainees as they continue to practice mixing concrete and performing the slump test.
2. Ensure that all safety rules are followed.

Class Break (10 minutes)

Module Examination (30 minutes)

1. Hand out Module Examinations to all trainees. Tell them they have 30 minutes to complete the exam.
2. Collect and grade the examinations.

Class Break (5 minutes)

Performance Profile Examination (45 minutes)

1. Have trainees perform the following tasks to the satisfaction of the instructor:
   - Mix a test batch of concrete.
   - Perform a slump test.
2. Ensure that all safety rules are followed.
3. Fill out Performance Profile Sheets for each trainee.

MODULE EXAMINATIONS AND ANSWER KEYS

The masters for Module Examinations and Answer Keys are found in the booklet entitled "Concrete Finishing Level One Module Examinations" included in this Instructor's Guide package. You may also reproduce the tests using the Windows-compatible software program which may be purchased separately from Prentice Hall. If you do not have the reproducible masters in your Instructor's Guide package, or would like to order the test software, please contact Prentice Hall Publishers at 1-800-922-0579.

Be sure to keep answer keys and tests in a safe area.

OVERHEAD TRANSPARENCIES

Masters for overhead transparencies are found on the following pages. Transparencies can be made by using clear acetate sheets in a photocopy machine. This can be done at any commercial copy shop, or by following this method on your office machine:

1. Use standard $8\frac{1}{2} \times 11$ clear acetate sheets, which can be obtained at office supply stores, stationery stores, or commercial copying centers.
2. Check the photocopier's manual to make sure it will take the thicker acetate sheets.
3. Remove the paper supply in the photocopying machine and replace it with a supply of acetate sheets.
4. Make photocopies from the masters on the following pages in the usual way.
   The material will be reproduced on the clear acetate sheets for use on any overhead projector.

TRANSPARENCY 1

TASK MODULE 23103, PROPERTIES OF CONCRETE

COURSE OBJECTIVES

Upon completion of this module, the trainee will be able to:

1. **Describe the properties of concrete.**
2. **Explain how the properties of concrete are used in construction.**
3. **Determine how the ingredients of concrete influence mix, placement, finishing, durability, and performance.**
4. **Describe quality-control tests on concrete ingredients, fresh concrete, and hardened concrete.**
5. **Mix a test batch of concrete.**
6. **Perform a slump test.**

TRANSPARENCY 2

TASK MODULE 23103, PROPERTIES OF CONCRETE

PERFORMANCE PROFILE TASKS

1. Mix a test batch of concrete.
2. Perform a slump test.

TRANSPARENCY 3

FIGURE 1

TASK MODULE 23103, PROPERTIES OF CONCRETE

A CUBIC FOOT OF TYPICAL CONCRETE

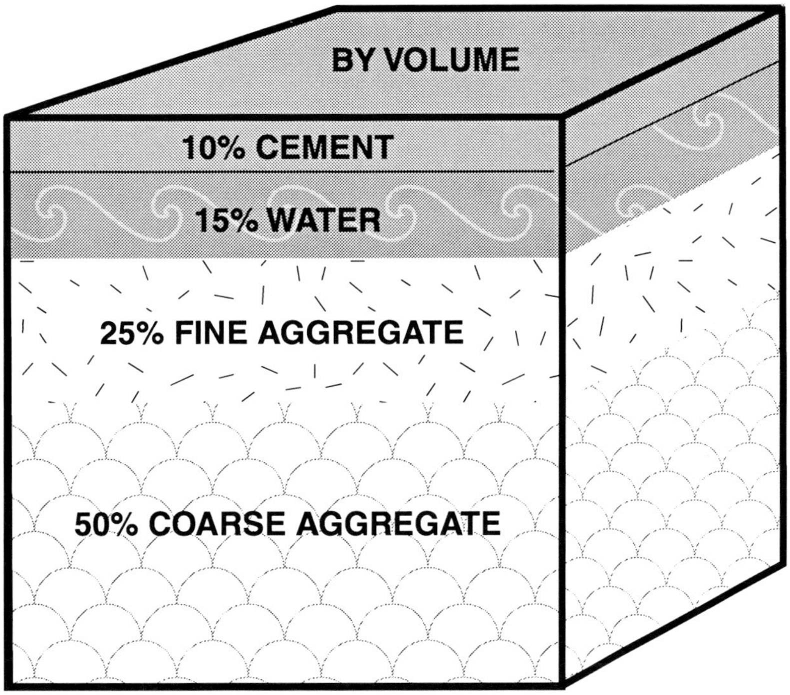

103F01.EPS

TRANSPARENCY 4

FIGURE 2

TASK MODULE 23103, PROPERTIES OF CONCRETE

CHARACTERISTICS AND TESTS OF AGGREGATES

CHARACTERISTIC	SIGNIFICANCE IN CONCRETE	TEST METHOD
Soundness	Strength, durability, appearance	ASTM C 88
Chemical stability	Alkali-silica reaction, popouts, disintegration, appearance	ASTM C 227, C 289 C 295, or C 342
Abrasion resistance	Wear resistance of floors, hardness	ASTM C 131 or C 535
Grading, or sieve analysis	Workability, density, economy, shrinkage	ASTM C 136
Maximum size of aggregate	Economy, shrinkage, density, strength	Inspection
Percentage of crushed particles	Workability, economy, strength	Separate and count particles
Particle shape	Workability, economy, shrinkage, strength	Inspection
Surface texture	Bond, strength, durability	Inspection
Specific gravity	Durability, density, need for mix computations	ASTM C 127 or C 128
Absorption	Durability, need for mix computations and control	ASTM C 127 or C 128
Moisture content	Need for mix computations and control	ASTM C 70 or C 566 or other

103F02.EPS

TRANSPARENCY 5

TASK MODULE 23103, PROPERTIES OF CONCRETE

CEMENT TYPES

PORTLAND CEMENTS

- Type I, Normal, general purpose
- Type II, Moderate Sulfate Resistant, for sulfates in groundwater
- Type III, High Early Strength, for cold weather or precasting plans
- Type IV, Low Heat of Hydration, for massive structures
- Type V, High Sulfate Resistant, for heavy sulfates in groundwater
- Type IA, IIA, IIIA, air-entrained admixture added

BLENDED CEMENTS

- Type IS, blast furnace slag
- Type IP, usually fly ash

TRANSPARENCY 6

TASK MODULE 23103, PROPERTIES OF CONCRETE

CONCRETE ADMIXTURES BY CLASSIFICATION

Type of Admixture	Desired Effect
Accelerators (ASTM C 494, Type C)	Accelerate setting and early-strength development
Air detrainers	Decrease air content
Air-entraining admixtures (ASTM C 260)	Improve durability in environments of freeze-thaw, deicers, sulfate, and alkali productivity. Improve workability
Alkali-reactivity reducers	Reduce alkali-reactivity expansion
Bonding Admixtures	Increase bond strength
Coloring Agents	Colored concrete
Corrosion inhibitors	Reduce steel corrosion activity in a chloride environment
Dampproofing admixtures	Retard moisture penetration into dry concrete.
Finely divided mineral admixtures	
Cementitious	Hydraulic properties. Partial cement replacement
Pozzolans	Pozzolanic activity. Improve workability, plasticity, sulfate resistance; reduce alkali reactivity, permeability, heat of hydration. Partial cement replacement. Filler
Pozzolanic and cementitious	Same as cementitious and pozzolan categories
Nominally inert	Improve workability. Filler
Fungicides, germicides, and insecticides	Inhibit or control bacterial and fungal growth
Gas formers	Cause expansion before setting

IG-TR6.EPS

TRANSPARENCY 7

TASK MODULE 23103, PROPERTIES OF CONCRETE

CONCRETE ADMIXTURES BY CLASSIFICATION (Cont.)

Type of Admixture	Desired Effect
Grouting agents	Adjust grout properties for specific applications
Permeability reducers	Decrease permeability
Pumping aids	Improve pumpability
Retarders (ASTM 494, Type B)	Retard setting time
Superplasticizers (ASTM C 1017, Type 1)	Flowing concrete Reduce water-cement ratio
Superplasticizers and retarder (ASTM C 1017, Type 2)	Flowing concrete with retarded set Reduce water
Water reducer (ASTM C 494, Type A)	Reduce water demand at least 5%
Water reducer and accelerator (ASTM C 494, Type E)	Reduce water (minimum 5%) and accelerate set
Water reducer and retarder (ASTM C 494, Type D)	Reduce water (minimum 5%) and retard set
Water reducer -- high range (ASTM C 494, Type F)	Reduce water demand (minimum 12%)
Water reducer--high range--and retarder (ASTM C 494, Type G)	Reduce water demand (minimum 12%) and retard set
Workability agents	Improve workability

IG-TR07.EPS

TRANSPARENCY 8

FIGURE 4

TASK MODULE 23103, PROPERTIES OF CONCRETE

FLY ASH PRODUCTION

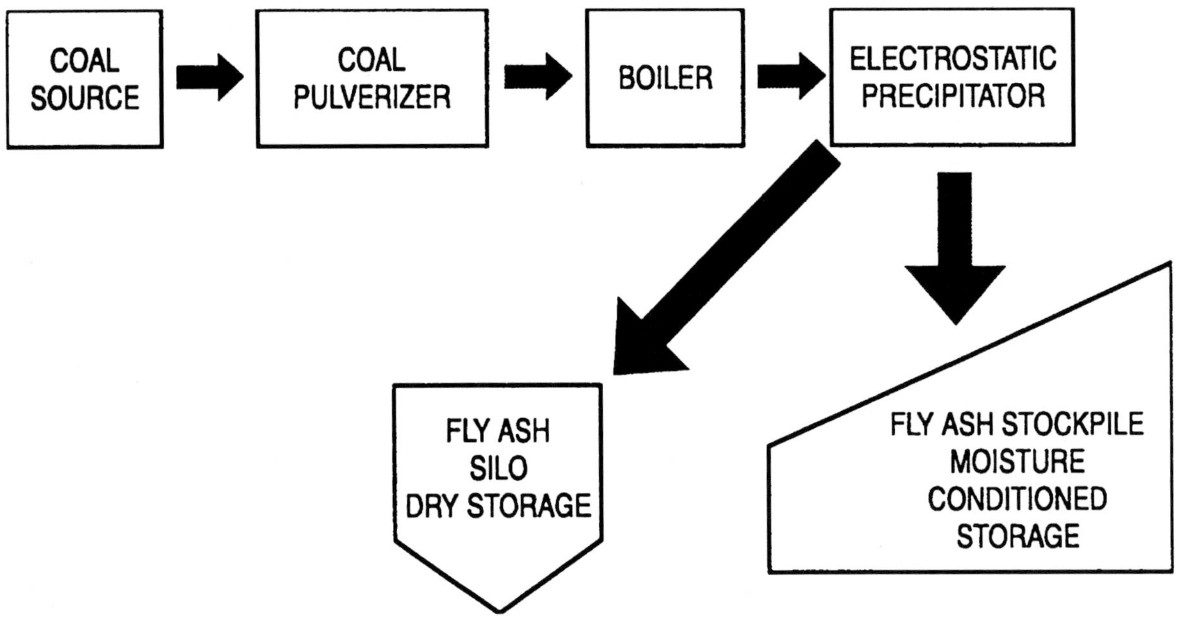

103F04.TIF

TRANSPARENCY 9

FIGURE 5

TASK MODULE 23103, PROPERTIES OF CONCRETE

EFFECTS OF ENTRAINED AIR ON CONCRETE PROPERTIES

Properties	Effect
Abrasion	Little effect; increased strength increases abrasion resistance
Absorption	Little effect
Alkali-silica reactivity	Expansion decreases with increased air
Bleeding	Reduced significantly
Bond to steel	Decreased
Compressive strength	Reduced approximately 2% to 6% per percentage point increase in air; harsh or lean mixes may gain strength
Creep	Little effect
Deicer scaling	Significantly reduced
Fatigue	Little effect
Flexural strength	Reduced approximately 2% to 4% per percentage point increase in air
Freeze-thaw resistance	Significantly improved resistance to water-saturated freeze-thaw deterioration
Heat of hydration	No significant effect
Modulus or elasticity (static)	Decreases with increased air approximately 105,000 to 200,000 psi per percentage point of air
Permeability	Little effect; reduced water-cement ratio reduces permeability

IG-TR9.EPS

TRANSPARENCY 10

FIGURE 5

TASK MODULE 23103, PROPERTIES OF CONCRETE

EFFECTS OF ENTRAINED AIR ON CONCRETE PROPERTIES (Cont.)

Properties	Effect
Scaling	Significantly reduced
Shrinkage (drying)	Little effect
Slump	Increases with air approximately 1 inch per 1/2 to 1 percentage point of air
Specific heat	No effect
Sulfate resistance	Significantly improved
Temperature of wet concrete	No effect
Thermal conductivity	Decreases 1% to 3% per percentage point increase of air
Thermal diffusivity	Decreases 1.6% per percentage point increase of air
Unit weight	Decreases with increased air
Water demand of wet concrete for equal slump	Decreases with increased air; approximately 5 to 10 pounds per cubic yard per percentage point of air
Watertightness	Increases slightly; reduced water-cement ratio increases watertightness
Workability	Increases with increased air

Note: The table information may not apply to all situations.

IG-TR10.EPS

TRANSPARENCY 11

FIGURE 10

TASK MODULE 23103, PROPERTIES OF CONCRETE

CONCRETE PROPORTIONS BY VOLUME

Aggregate Size, Inches	With or Without Air Entrainment			With Air Entrainment	Without Air Entrainment
	Cement	Water *	Coarse Aggregate	Sand	Sand
3/8	1	1/2	1-1/2	2-1/4	2-1/2
1/2	1	1/2	2	2-1/4	2-1/2
3/4	1	1/2	2-1/2	2-1/4	2-1/2
1	1	1/2	2-3/4	2-1/4	2-1/2
1-1/2	1	1/2	3	2-1/4	2-1/2

Unit of measure is either cubic foot or bucket.

*** 7.48 gallons of water equal 1 cubic foot.**

103F10.EPS

TRANSPARENCY 12

FIGURE 13

TASK MODULE 23103, PROPERTIES OF CONCRETE

TYPES OF STRENGTH

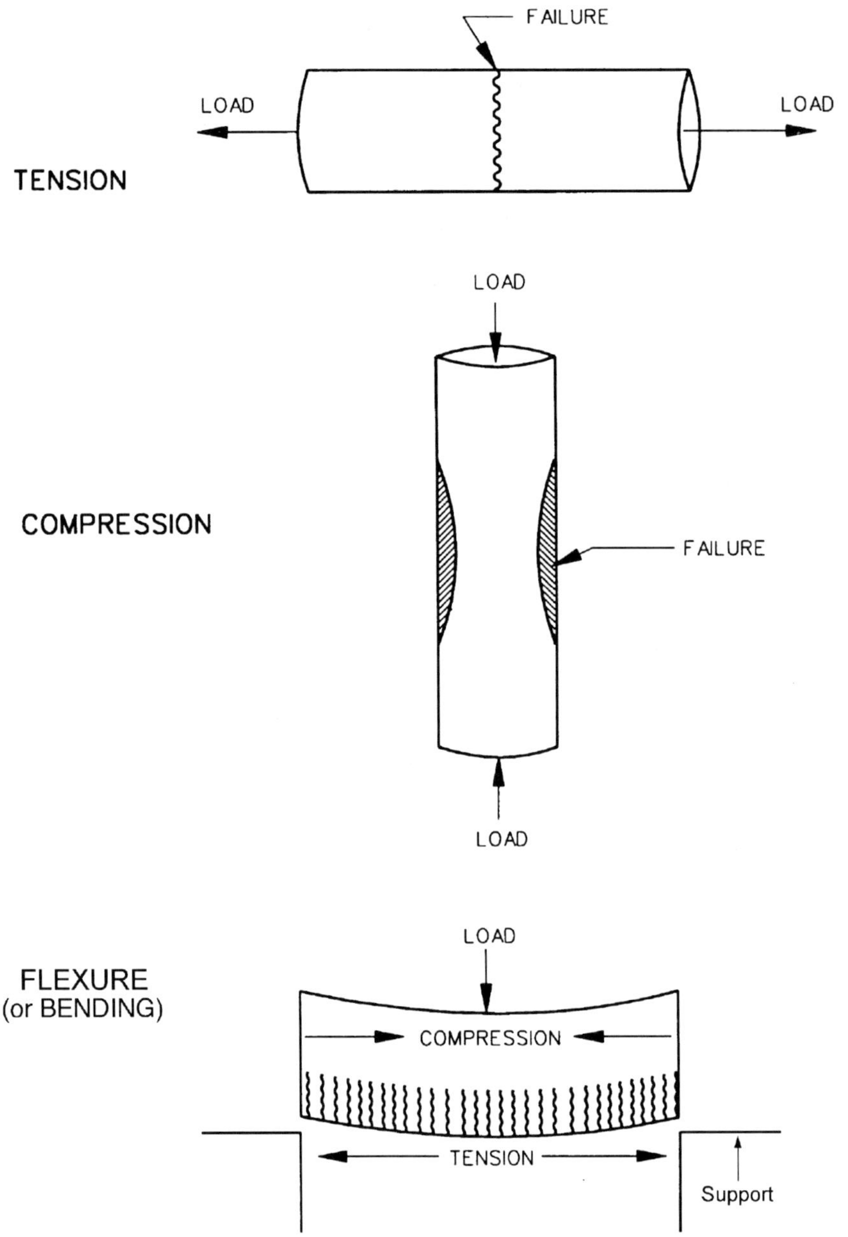

103F13.TIF

TRANSPARENCY 13

FIGURE 14

TASK MODULE 23103, PROPERTIES OF CONCRETE

CONCRETE AGE-STRENGTH RELATIONSHIPS

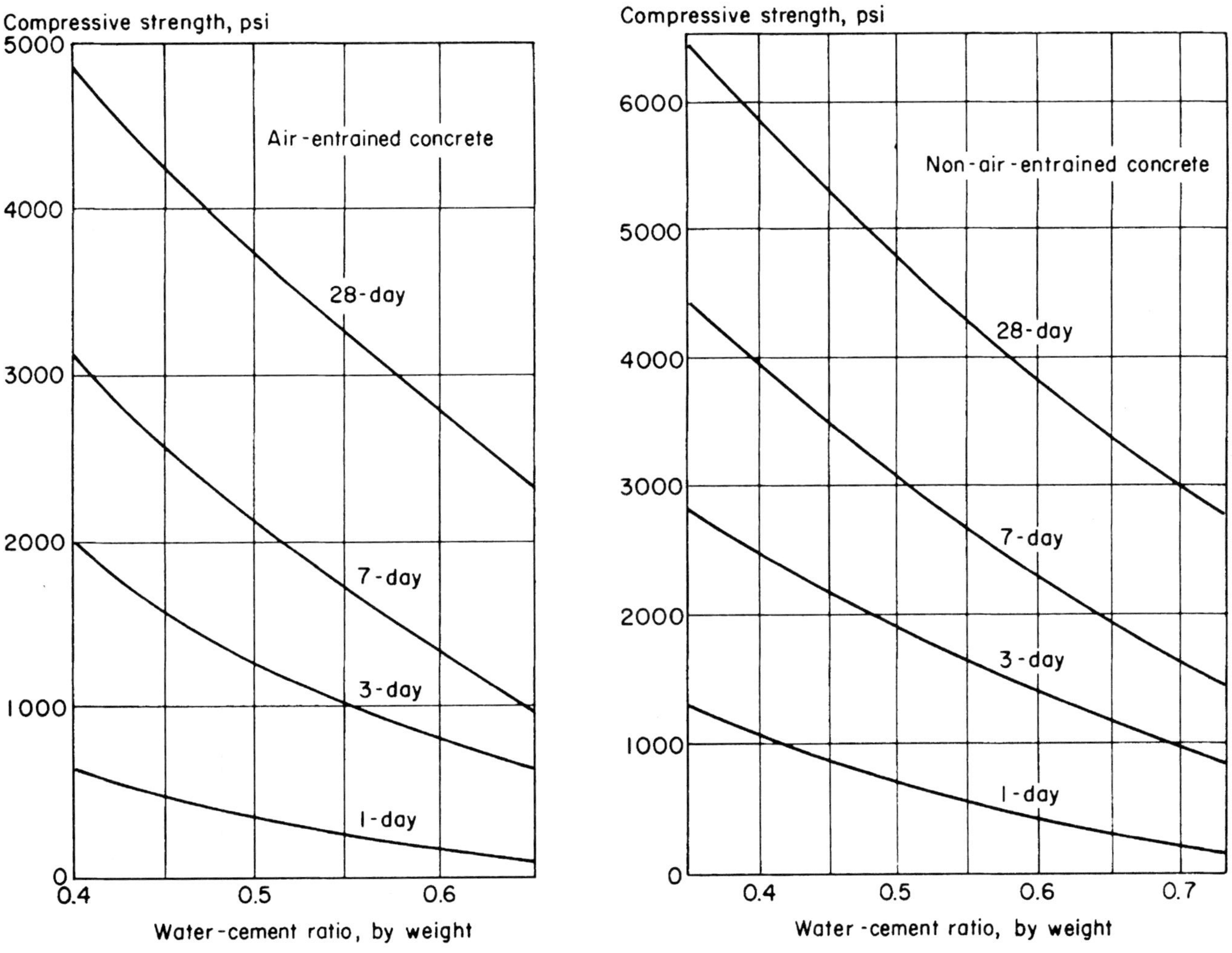

103F14.TIF

TRANSPARENCY 14

FIGURE 16

TASK MODULE 23103, PROPERTIES OF CONCRETE

VARIOUS TYPES OF REINFORCING BARS

BAR SIZE		NOMINAL DIMENSIONS *					
		DIAMETER		CROSS-SECTIONAL AREA		WEIGHT	
SI Metric	US Cust.	mm	inches	mm^2	$inches^2$	kg/m	lbs/ft
#10		9.5		71		0.560	
#13	#4	12.7	0.500	129	0.20	0.944	0.668
#16	#5	15.9	0.625	199	0.31	1.552	1.043
#19	#6	19.1	0.750	284	0.44	2.235	1.502
#22	#7	22.2	0.875	387	0.60	3.042	2.044
#25	#8	25.4	1.000	510	0.79	3.973	2.670
#29	#9	28.7	1.128	645	1.00	5.060	3.400
#32	#10	32.3	1.270	819	1.27	6.404	4.303
#36	#11	35.8	1.410	1006	1.56	7.907	5.313
#43	#14	43.0	1.693	1452	2.25	11.380	7.650
#57	#18	57.3	2.257	2591	4.00	20.240	13.600

103F16.EPS

TRANSPARENCY 15

FIGURE 21

TASK MODULE 23103, PROPERTIES OF CONCRETE

EQUIPMENT FOR SLUMP TEST

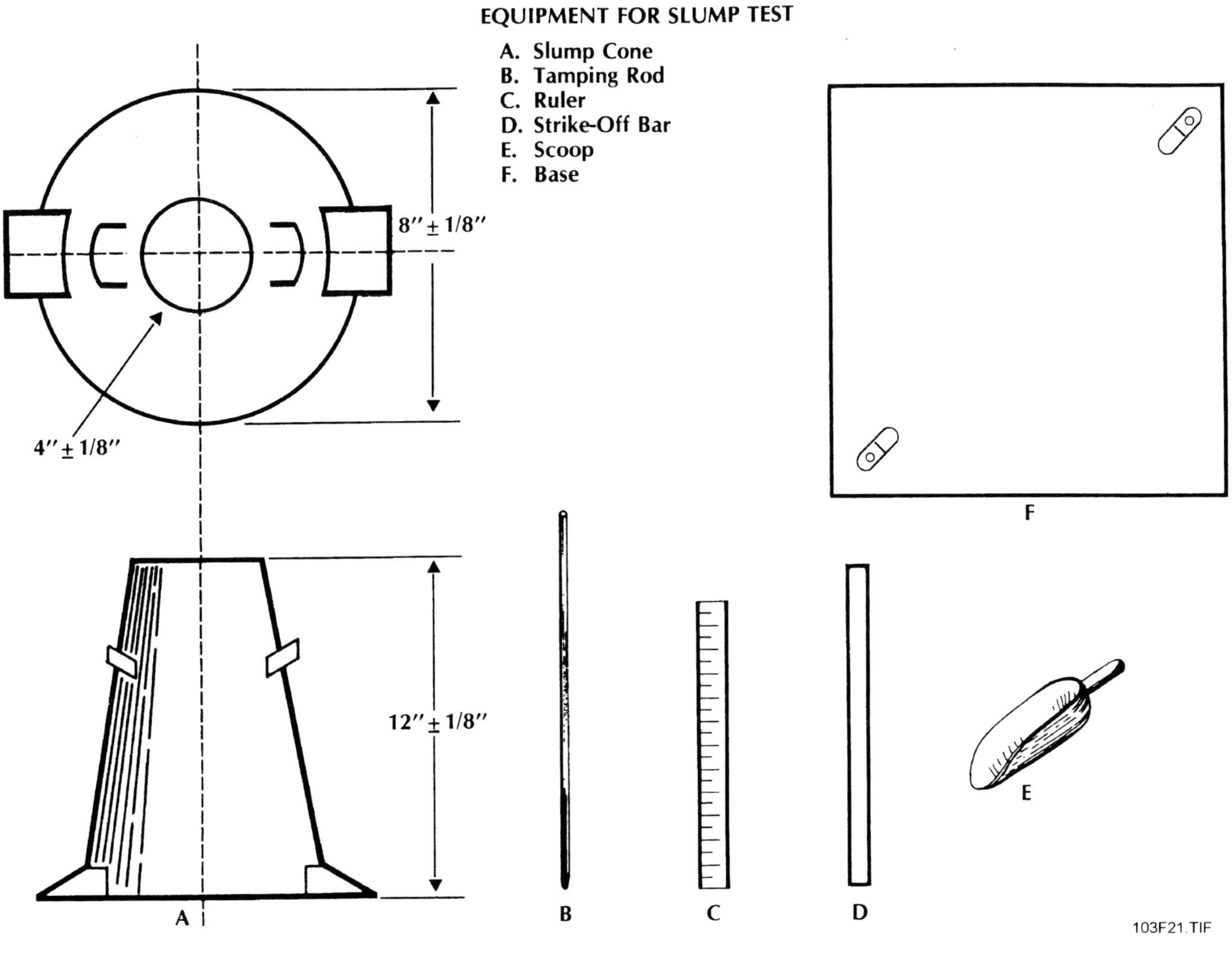

TRANSPARENCY 16

FIGURE 23

TASK MODULE 23103, PROPERTIES OF CONCRETE

SLUMP MEASUREMENT

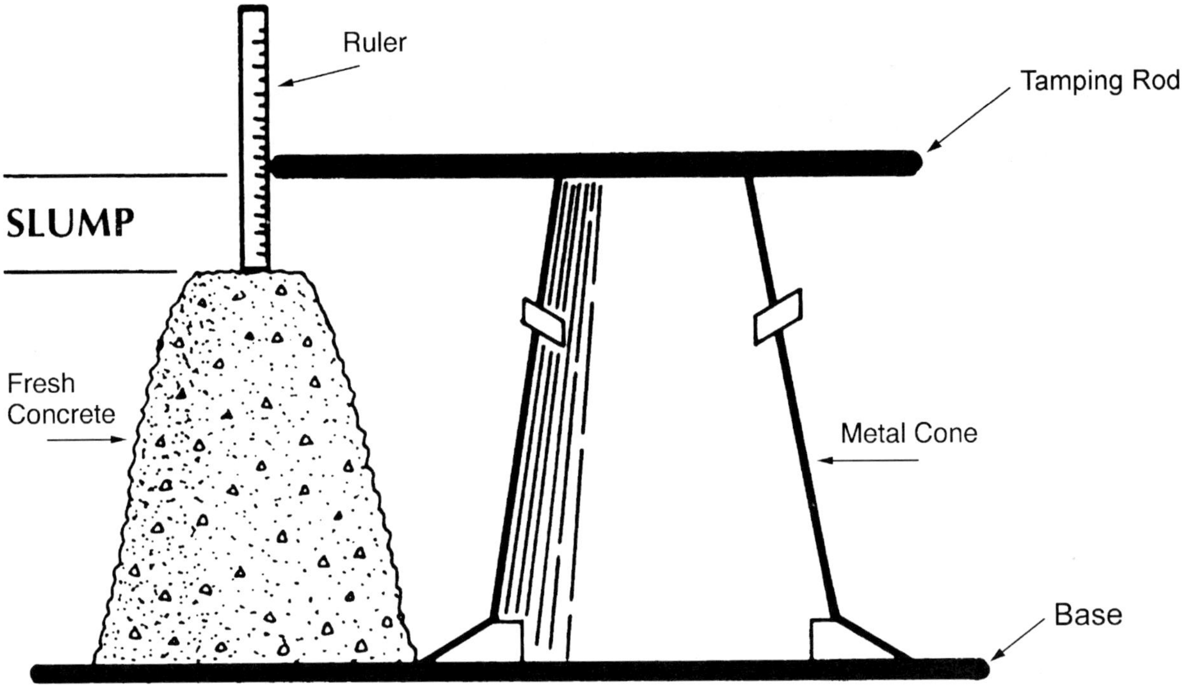

103F23.TIF

NOTE ON PERFORMANCE PROFILE TESTING

Performance Profiles are included in this Instructor's Guide in a format that can be easily photocopied for each trainee. The Profiles measure trainee competency in the tasks taught in this module.

Please note the number of tasks to be tested while teaching this module. Each trainee should be tested on the tasks listed in the Performance Profile. Before the performance testing, the instructor should brief the trainees on:

- test objectives and criteria,
- safety precautions, and
- procedures for each task to be tested.

The instructor administering the performance testing should also do the following:

- ensure that all of the needed equipment is available and operating properly;
- set up the testing stations;
- organize and administer the test in a way that allows for optimal performance;
- complete the Performance Profile Sheet for each trainee by assigning a score for each listed task;
- monitor adherence to all safety regulations and precautions;
- provide adequate supervision to prevent injuries; and
- take immediate and effective action to remedy any emergency.

ACCREDITATION TESTING

If this Performance Profile Testing is done in the National Center for Construction Education and Research Standardized Craft Training Program, the following conditions must be met:

1. The Craft Instructor must hold valid NCCER instructor certification.
2. The training must be delivered through a Training Program Sponsor recognized by the NCCER.
3. For every module, the specific performance testing must be completed to the satisfaction of the instructor.
4. The results of the testing must be recorded on the Craft Training Report Form. This form must be provided to the local Training Program Sponsor to be forwarded to the NCCER National Craft Training Registry.

Craft: Concrete Finishing	WHEELS
Task Module Number: 23103	OF
Task Module Title: Properties of Concrete	LEARNING

TRAINEE NAME: ______________________________

TRAINEE SOCIAL SECURITY NUMBER: ______________________________

CLASS: ______________________________

TRAINING PROGRAM SPONSOR: ______________________________

____

INSTRUCTOR: ______________________________

Rating Levels: 1. Passed: performed task.

2. Failed: did not perform task.

Recognition: When testing for the NCCER Standardized Craft Training Program, be sure to record Performance Profile testing results on Craft Training Report Form 200 and submit the results to the Training Program Sponsor.

Objective	TASK	RATING
1, 3, 5	1. Mix a test batch of concrete.	
4, 6	2. Perform a slump test.	

NCCER CRAFT TRAINING USER UPDATES

The NCCER makes every effort to keep these manuals up-to-date and free of technical errors. We appreciate your help in this process. If you have an idea for improving this manual, or if you find an error, a typographical mistake, or an inaccuracy in the NCCER's Craft Training Manuals, please write us, using this form or a photocopy. Be sure to include the exact module number, page number, a description of the problem, and the correction, if possible. Your input will be brought to the attention of the Technical Review Committee. Thank you for your assistance.

Instructors – If you found that additional materials were necessary in order to teach this module effectively, please let us know so that we may include them in the Equipment/Materials list in the Instructor's Guide.

Write: Curriculum and Revision Department
National Center for Construction Education and Research
P.O. Box 141104
Gainesville, FL 32614-1104

Fax: 352-334-0932

Craft ______________________ Module Name ______________________

Module Number ______________________ Page Number(s) ______________________

Description of Problem

____

____

____

____

(Optional) Correction of Problem

____

____

____

(Optional) Your Name and Address

____

____

____

notes

Tools and Equipment

Module 23104

NATIONAL
CENTER FOR
CONSTRUCTION
EDUCATION AND
RESEARCH

TOOLS AND EQUIPMENT

OBJECTIVES

Upon completion of this module, the trainee will be able to:

1. Name the tools used in placing and finishing concrete.
2. Name the power equipment used in placing and finishing concrete.
3. Describe how each tool is used.
4. Describe how the power equipment is used.
5. Associate trade terms with the appropriate tools and equipment.

Note to the Instructor

Before teaching this module, you should review the details in this Instructor's Guide for Equipment and Supplies, Testing, and the suggested Teaching Sequence. Be sure to allow ample time to prepare your own training plan or lesson plan and to gather all required equipment and materials.

Required Equipment and Materials

The following are required for instruction using this module:

Equipment
Overhead projector and screen
Whiteboard/chalkboard
Appropriate Personal Protective Equipment
Steel tape
Engineer's rule
Carpenter's level
Builder's level
Straightedge
Rake
Come-along
Bullfloat
Shovel
Highway straightedge
Darby
Rubber float
Finishing trowel
Walking trowel
Pointing trowel
Hand-held jointer
Basic edger
Pair edgers
Radius edger
Finishing broom
Water broom
Hand stone
Hand-held concrete vibrator
Gasoline-powered concrete saw
Electric grinder
Walk behind power trowel

Materials
Trainee Task Module
Transparencies
Markers/chalk
Pencils and paper
Module Examinations
Performance Profile Sheets

HOW TO USE THIS INSTRUCTOR'S GUIDE

For each 2½ hour class session in this Instructor's Guide, the basic Presentation Sequence is as follows:

Introduction/Overview

Classroom, and/or Demonstration, and/or Laboratory

Class Break

Classroom, and/or Demonstration, and/or Laboratory

Summary

Suggested time periods for classroom sessions are included throughout this Instructor's Guide. These time periods should be adapted to meet local conditions and training requirements.

Each class session is presented with two columns of information. On the left side of the page, a narrow column provides suggested trainee and instructor actions, icons to call your attention to material, safety, audiovisual, or testing requirements, and space for your notes. The right-hand column provides the outline of the suggested presentation for each class session.

In this Instructor's Guide, the terms classroom, demonstration, and laboratory are defined and used as follows.

Classroom: Sessions are designed for lectures, group discussions, coaching, and additional activities. Trainees should be encouraged to actively participate.

Demonstration: Instructors will demonstrate all procedures before trainees attempt them. Instructors should make sure that trainees can point out all safety procedures during demonstrations to be assured of the proper use of equipment by trainees.

Laboratory: Instructors will facilitate all laboratory activities, coach trainees as they practice the procedures, monitor trainee progress, and provide feedback. The instructor will make sure that safety rules are followed at all times and that protective equipment is worn.

NCCER Standardized Craft Training Programs

The National Center for Construction Education and Research (NCCER) provides a standardized national program of accredited craft training. Key features of the program include instructor certification, competency-based training, and performance testing. The program provides trainees, instructors, and companies with a standard form of recognition through a National Craft Training Registry. The program is described in full in the *Guidelines for Accreditation*, published by the NCCER. For more information on standardized craft training, contact the NCCER at P.O. Box 141104, Gainesville, FL 32614-1104; or call 352-334-0911.

MODULE OVERVIEW

This module introduces the Concrete Finishing trainee to the tools and equipment used in placing and finishing concrete.

Prerequisites

Please see the Course Map. Prior to training with this module, it is recommended that the trainee shall have successfully completed the following modules:

Core Curricula

Concrete Finishing Level 1, Modules 23101 through 23103

Safety Considerations

Ensure that the trainees are equipped with Appropriate Personal Protective Equipment.

Teaching Time for This Module

Approximately 7½ hours or three sessions of training time are suggested to cover *Tools And Equipment*. The training class session is a suggested 2½ hour time period, which includes at least one break. **You will need to adjust the time required for hands-on activities and testing based on your class size and resources.** All time periods for this module are suggested and you will need to adapt the suggested lesson plan to meet your local conditions.

Suggested Teaching Sequence — Three 2½-Hour Sessions

Adjust your class times based on class size and resources.

Session	Topic	Trainee Module Section(s)
1	Introduction; Placing and Finishing Tools; Equipment	1.0.0 – 3.17.0
2	Laboratory/Site Visit	1.0.0 – 3.17.0
3	Equipment; Module Examination Performance Profile Test	3.13.0 – 3.17.0

Optional References for Advanced Study

This module is intended to present thorough resources for task training. The following reference works are suggested for both instructors and motivated trainees interested in further study. These are optional materials for continued education rather than for task training.

Concrete Construction and Estimating, Craig Avery, Craftsman Book Company, Carlsbad, CA, 1991.

Concrete Manual, International Conference of Building Officials, Whittier, CA, 1992.

Concrete Parking Lots: Eight Steps to Quality Construction, Videotape, Portland Cement Association, Skokie, IL

Concrete Technology, Third Edition, George R. White, Delmar Publishers, Inc., Albany, NY, 1991.

Design and Control of Concrete Mixtures, Thirteenth Edition, Portland Cement Association, Skokie, IL, 1988.

Finishing Concrete Flatwork, Videotape 32:00 minutes, Portland Cement Association, Skokie, IL, 1991.

Guide for Concrete Floor and Slab Construction, ACI 302.1R-96, American Concrete Institute, Farmington Hills, MI, 1997.

The Contractor's Guide to Quality Construction, American Concrete Institute, Farmington Hills, MI, 1997.

COURSE MAP

This course map shows all of the modules in the first level of the Concrete Finishing curricula. The suggested training order begins at the bottom and proceeds up. Skill levels increase as a trainee advances on the course map. The training order may be adjusted by the local Training Program Sponsor.

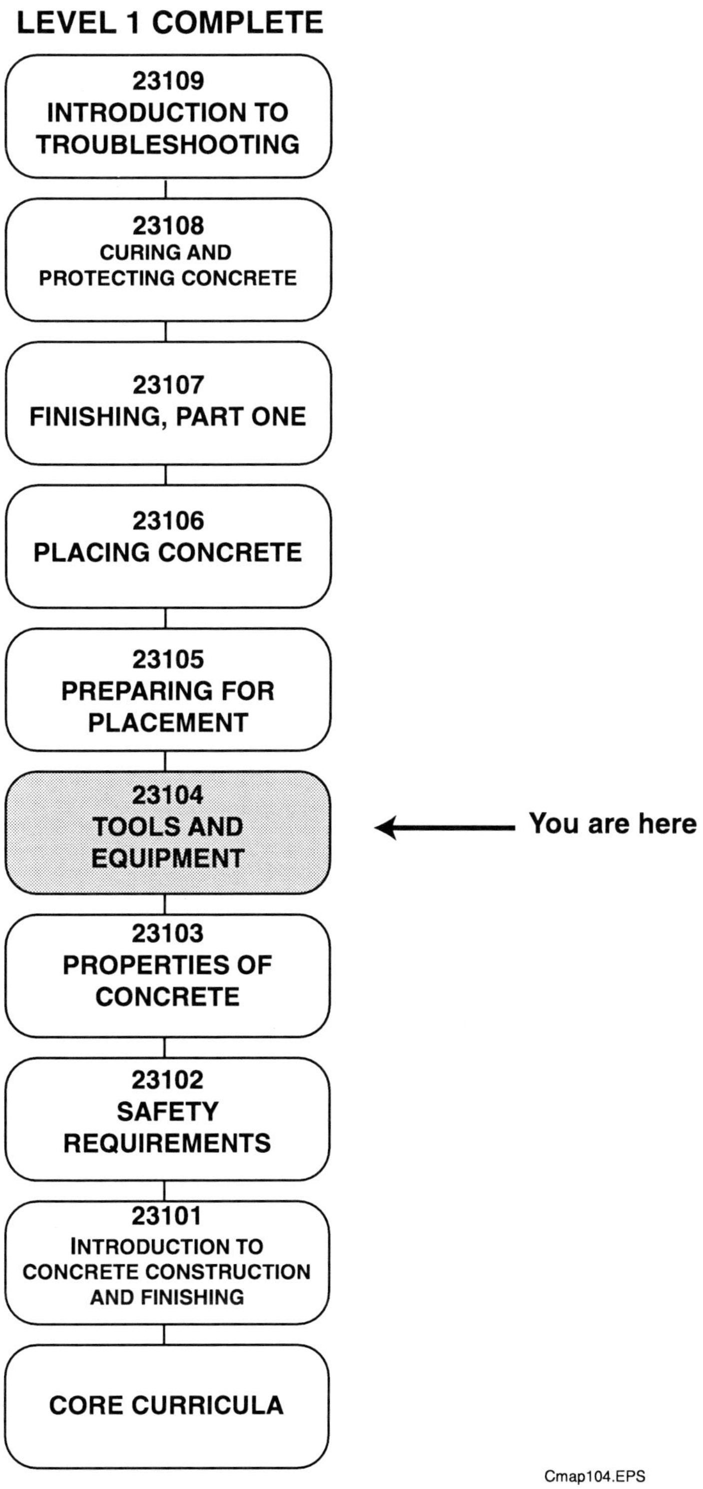

Cmap104.EPS

Ensure that all trainees have brought their Trainee Modules to class.

Equipment and Materials

Trainee Task Module
Overhead projector and screen
Whiteboard/chalkboard
Markers/chalk
Pencils and paper
Transparencies
Appropriate Personal Protective Equipment
Steel tape
Engineer's rule
Carpenter's level
Builder's level
Straightedge
Rake
Come-along
Bullfloat
Shovel
Highway straightedge
Darby
Rubber float
Finishing trowel
Walking trowel
Pointing trowel
Hand-held jointer
Basic edger
Pair edgers
Radius edger
Finishing broom
Water broom
Hand stone
Hand-held concrete vibrator

Time Schedule (150 minutes)

15 Overview
45 Classroom/Demonstration
10 Class Break
65 Classroom/Demonstration
15 Summary

Overview (15 minutes)

Show Transparency 1, Course Objectives.

Show Transparency 2, Performance Profile Tasks.

1. Take attendance and perform other administrative tasks.
2. Outline briefly the topics to be covered. Explain to the trainees that they must pass the Module Examination with a grade of at least 70 percent. Also, explain that they must pass the Performance Profile Examination by performing each task to the level of acceptance criteria.
3. Provide a brief explanation of the acceptance criteria that will be used in the Performance Profile Examination.

Classroom/Demonstration (45 minutes)

Ensure that all hand tools demonstrated are safe and in good working order.

1. Review the trade terms introduced in this module and explain them.
2. Identify and describe each hand tool included in the Trainee Task Module. Explain the use of the tool, how it is handled, safety considerations, and maintenance requirements.
3. Identify and discuss measuring tools: rules, tapes, and levels. Pass the tools through the class after you describe them.
4. Identify and discuss placing tools: rakes, come-alongs, and shovels. Pass the tools through the class.
5. Identify and discuss leveling tools: screeds and straightedges. Pass them through the class.
6. Identify and discuss finishing tools: hand floats, bullfloats, darbys, and highway straightedges. Pass them through the class.

Class Break (10 minutes)

Classroom/Demonstration (65 minutes)

1. Identify and discuss groovers and edgers. Pass them through the class.
2. Identify and discuss hand trowels, fresno trowels, and pointing and margin trowels. Pass them through the class.
3. Identify and discuss brushes, brooms, and rubbing stones. Pass them through the class.
4. Identify and discuss any other hand tools that may be used by your company.

Show Transparency 3, Vibrator Size and Placing (Figure 52, Trainee Module).

5. Identify and discuss concrete vibrators. Pass them through the class.
6. Identify and discuss compactors and rollers.

Summary (15 minutes)

1. Review the materials covered in this session.
2. Answer trainee questions about the material covered.
3. Explain that the next session will be a site visit to a construction project where many of the tools and equipment described in this module are being used. Emphasize the need to understand the operation and safe use of larger pieces of equipment such as power buggies, power floats and trowels, and ready-mix trucks.

SESSION 2 OF 3 SESSIONS

Equipment and Materials

Pencils and paper
Appropriate Personal Protective Equipment

Ensure that all trainees have brought writing materials to the site.

Time Schedule (150 minutes)

15 Overview
95 Laboratory/Site Visit
10 Class Break
30 Summary

Overview (15 minutes)

1. Take attendance and perform other administrative tasks.
2. Discuss taking notes and assign specific responsibilities.
3. Impress on trainees the importance of:
   - staying alert
   - staying with the group
   - obeying instructions

Laboratory/Site Visit (95 minutes)

Make sure that arrangements have been made for trainees to tour a construction site.

Coordinate visit with appropriate site personnel.

Ensure that sufficient transportation is available.

1. On the site tour, point out the tools and equipment in use.
2. Point out construction stages and other material covered in previous modules.
3. Answer questions from the trainees.

Class Break (10 minutes)

Summary (30 minutes)

1. Review the events and items observed.
   - Ask trainees to list tools and equipment they saw in use.
   - Ask trainees to list tools and equipment they did not see in use.
2. Give assignments for the next session.
3. Tell trainees to complete the Review/Practice Questions in the back of the Trainee Task Module.
4. Explain that the next session is the last session for this module and that there will be a Module Examination and Performance Profile Test.

SESSION 3 OF 3 SESSIONS

Ensure that all trainees have brought their Trainee Modules to class.

Equipment and Materials

Trainee Task Module
Overhead projector and screen
Whiteboard/chalkboard
Markers/chalk
Transparencies
Pencils and paper
Appropriate Personal Protective Equipment
Gasoline-powered concrete saw
Walk behind power trowel
Electric grinder
Module Examinations
Performance Profile Sheets

Time Schedule (150 minutes)

5 Overview
45 Classroom/Demonstration
10 Class Break
30 Module Examination
10 Class Break
50 Performance Profile Examination

Show Transparency 1, Course Objectives.

Show Transparency 2, Performance Profile Tasks.

Overview (5 minutes)

1. Take attendance and perform other administrative tasks.
2. Outline briefly the topics to be covered. Explain to the trainees that they must pass the Module Examination with a grade of at least 70 percent. Also, explain that they must pass the Performance Profile Examination by performing each task to the level of acceptance criteria.
3. Provide a brief explanation of the acceptance criteria that will be used in the Performance Profile Examination.

Classroom/Demonstration (45 minutes)

1. Describe the different types of concrete saws that can be used to cut and shape concrete. Point out the need to check the equipment before use. Emphasize safe use of equipment and the need for Appropriate Personal Protective Equipment.

2. Describe the different types of power floats and their uses.
3. Describe the different types of power trowels and their uses.
4. Describe the different types of concrete grinding and milling machines found on a typical job.
5. Go over answers to the Review/Practice Questions with trainees. Answer any questions the trainees may have.

Class Break (10 minutes)

Module Examination (30 minutes)

1. Hand out Module Examinations to all trainees. Tell them they have 30 minutes to complete the exam.
2. Collect and grade the examinations.

Class Break (10 minutes)

Performance Profile Examination (50 minutes)

1. Have trainees perform the following tasks to the satisfaction of the instructor:
   - Identify selected hand tools and their components.
   - Identify selected power tools and equipment.
2. Fill out Performance Profile Sheets for each trainee.

MODULE EXAMINATIONS AND ANSWER KEYS

The masters for Module Examinations and Answer Keys are found in the booklet entitled "Concrete Finishing Level One Module Examinations" included in this Instructor's Guide package. You may also reproduce the tests using the Windows-compatible software program which may be purchased separately from Prentice Hall. If you do not have the reproducible masters in your Instructor's Guide package, or would like to order the test software, please contact Prentice Hall Publishers at 1-800-922-0579.

Be sure to keep answer keys and tests in a safe area.

OVERHEAD TRANSPARENCIES

Masters for overhead transparencies are found on the following pages. Transparencies can be made by using clear acetate sheets in a photocopy machine. This can be done at any commercial copy shop, or by following this method on your office machine:

1. Use standard $8\frac{1}{2}$ x 11 clear acetate sheets, which can be obtained at office supply stores, stationery stores, or commercial copying centers.
2. Check the photocopier's manual to make sure it will take the thicker acetate sheets.
3. Remove the paper supply in the photocopying machine and replace it with a supply of acetate sheets.
4. Make photocopies from the masters on the following pages in the usual way.
   The material will be reproduced on the clear acetate sheets for use on any overhead projector.

TRANSPARENCY 1

TASK MODULE 23104, TOOLS AND EQUIPMENT

COURSE OBJECTIVES

Upon completion of this module, the trainee will be able to:

1. **Name the tools used in placing and finishing concrete.**
2. **Name the power equipment used in placing and finishing concrete.**
3. **Describe how each tool is used.**
4. **Describe how the power equipment is used.**
5. **Associate trade terms with the appropriate tools and equipment.**

TRANSPARENCY 2

TASK MODULE 23104, TOOLS AND EQUIPMENT

PERFORMANCE PROFILE TASKS

1. Identify selected hand tools and their components.

2. Identify selected power tools and equipment.

TRANSPARENCY 3

FIGURE 52

TASK MODULE 23104, TOOLS AND EQUIPMENT

VIBRATOR SIZE AND PLACING

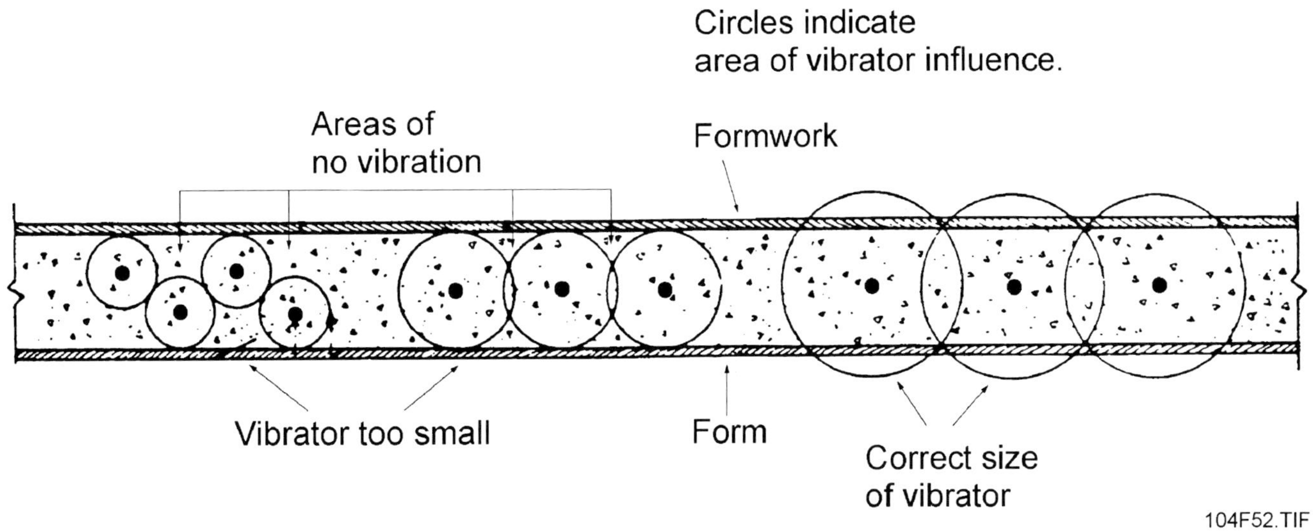

NOTE ON PERFORMANCE PROFILE TESTING

Performance Profiles are included in this Instructor's Guide in a format that can be easily photocopied for each trainee. The Profiles measure trainee competency in the tasks taught in this module.

Please note the number of tasks to be tested while teaching this module. Each trainee should be tested on the tasks listed in the Performance Profile. Before the performance testing, the instructor should brief the trainees on:

- test objectives and criteria,
- safety precautions, and
- procedures for each task to be tested.

The instructor administering the performance testing should also do the following:

- ensure that all of the needed equipment is available and operating properly;
- set up the testing stations;
- organize and administer the test in a way that allows for optimal performance;
- complete the Performance Profile Sheet for each trainee by assigning a score for each listed task;
- monitor adherence to all safety regulations and precautions;
- provide adequate supervision to prevent injuries; and
- take immediate and effective action to remedy any emergency.

ACCREDITATION TESTING

If this Performance Profile Testing is done in the National Center for Construction Education and Research Standardized Craft Training Program, the following conditions must be met:

1. The Craft Instructor must hold valid NCCER instructor certification.
2. The training must be delivered through a Training Program Sponsor recognized by the NCCER.
3. For every module, the specific performance testing must be completed to the satisfaction of the instructor.
4. The results of the testing must be recorded on the Craft Training Report Form. This form must be provided to the local Training Program Sponsor to be forwarded to the NCCER National Craft Training Registry.

Craft: Concrete Finishing	WHEELS
Task Module Number: 23104	OF
Task Module Title: Tools And Equipment	LEARNING

TRAINEE NAME: ______________________________

TRAINEE SOCIAL SECURITY NUMBER: ______________________________

CLASS: ______________________________

TRAINING PROGRAM SPONSOR: ______________________________

____

INSTRUCTOR: ______________________________

Rating Levels:
1. Passed: performed task.
2. Failed: did not perform task.

Recognition: When testing for the NCCER Standardized Craft Training Program, be sure to record Performance Profile testing results on Craft Training Report Form 200 and submit the results to the Training Program Sponsor.

Objective	TASK	RATING
1, 3	Identify selected hand tools and their components.	
2, 4	Identify selected power tools and equipment.	

NCCER CRAFT TRAINING USER UPDATES

The NCCER makes every effort to keep these manuals up-to-date and free of technical errors. We appreciate your help in this process. If you have an idea for improving this manual, or if you find an error, a typographical mistake, or an inaccuracy in the NCCER's Craft Training Manuals, please write us, using this form or a photocopy. Be sure to include the exact module number, page number, a description of the problem, and the correction, if possible. Your input will be brought to the attention of the Technical Review Committee. Thank you for your assistance.

Instructors – If you found that additional materials were necessary in order to teach this module effectively, please let us know so that we may include them in the Equipment/Materials list in the Instructor's Guide.

Write: Curriculum and Revision Department
National Center for Construction Education and Research
P.O. Box 141104
Gainesville, FL 32614-1104

Fax: 352-334-0932

Craft ________________ Module Name ________________

Module Number ________________ Page Number(s) ________________

Description of Problem

____

____

____

____

(Optional) Correction of Problem

____

____

____

(Optional) Your Name and Address

____

____

____

Preparing for Placement

Module 23105

Concrete Finishing Instructor's Guide Task Module 23105

PREPARING FOR PLACEMENT

NATIONAL
CENTER FOR
CONSTRUCTION
EDUCATION AND
RESEARCH

OBJECTIVES

Upon completion of this module, the trainee will be able to:

1. Describe basic site layout using levels and measuring tools.
2. Properly locate, grade, and build forms for horizontal placement.
3. Perform compaction activities on subgrades.
4. Describe various joints and where to locate them.
5. Describe various reinforcements and how to place them.
6. Describe information needed when ordering concrete.

Note to the Instructor

Before teaching this module, you should review the details in this Instructor's Guide for Equipment and Supplies, Testing, and the suggested Teaching Sequence. Be sure to allow ample time to prepare your own training plan or lesson plan and to gather all required equipment and materials.

Required Equipment and Materials

The following are required for instruction using this module:

Equipment
Overhead projector and screen
Whiteboard/chalkboard
Engineer's level
Laser level
Level rods
Hand level
Measuring tape
Sandbox(es)
Shovels
Come-alongs
Hammers
Appropriate Personal Protective Equipment

Materials
Trainee Task Module
Transparencies
Markers/chalk
Pencils and paper
Construction stakes
Batter boards
Dimensional lumber (2 × 4 and 2 × 6)
Nails
String (mason's line) and cutter
Grease crayon or marking pen
Reinforcing steel
Waterstop
Site plan
Module Examinations
Performance Profile Sheets

HOW TO USE THIS INSTRUCTOR'S GUIDE

For each 2½ hour class session in this Instructor's Guide, the basic Presentation Sequence is as follows:

Introduction/Overview

Classroom, and/or Demonstration, and/or Laboratory

Class Break

Classroom, and/or Demonstration, and/or Laboratory

Summary

Suggested time periods for classroom sessions are included throughout this Instructor's Guide. These time periods should be adapted to meet local conditions and training requirements.

Each class session is presented with two columns of information. On the left side of the page, a narrow column provides suggested trainee and instructor actions, icons to call your attention to material, safety, audiovisual, or testing requirements, and space for your notes. The right-hand column provides the outline of the suggested presentation for each class session.

In this Instructor's Guide, the terms classroom, demonstration, and laboratory are defined and used as follows:

Classroom: Sessions are designed for lectures, group discussions, coaching, and additional activities. Trainees should be encouraged to actively participate.

Demonstration: Instructors will demonstrate all procedures before trainees attempt them. Instructors should make sure that trainees can point out all safety procedures during demonstrations to be assured of the proper use of equipment by trainees.

Laboratory: Instructors will facilitate all laboratory activities, coach trainees as they practice the procedures, monitor trainee progress, and provide feedback. The instructor will make sure that safety rules are followed at all times and that protective equipment is worn.

NCCER Standardized Craft Training Programs

The National Center for Construction Education and Research (NCCER) provides a standardized national program of accredited craft training. Key features of the program include instructor certification, competency-based training, and performance testing. The program provides trainees, instructors, and companies with a standard form of recognition through a National Craft Training Registry. The program is described in full in the *Guidelines for Accreditation*, published by the NCCER. For more information on standardized craft training, contact the NCCER at P.O. Box 141104, Gainesville, FL 32614-1104; or call 352-334-0911.

MODULE OVERVIEW

This module provides training in the required activities and methods used in preparing a site for placing concrete. Topics include the processes from site excavation through ordering the concrete. Proper placement and inspection of formwork is emphasized.

Prerequisites

Please see the Course Map. Prior to training with this module, it is recommended that the trainee shall have successfully completed the following modules:

Core Curricula

Concrete Finishing Level 1, Modules 23101 through 23104

Safety Considerations

Ensure that the trainees are equipped with Appropriate Personal Protective Equipment.

Teaching Time for This Module

Approximately 12½ hours or five sessions of training time are suggested to cover *Preparing for Placement*. The training class session is a suggested 2½ hour time period, which includes at least one break. **You will need to adjust the time required for hands-on activities and testing based on your class size and resources.** All time periods for this module are suggested, and you will need to adapt the suggested lesson plan to meet your local conditions.

Suggested Teaching Sequence — Five 2½-Hour Sessions

Adjust your class times based on class size and resources.

Session	Topic	Trainee Module Section(s)
1	Introduction; Grade and Subgrade Preparation	1.0.0 – 2.5.2
2	Formwork; Surveying/Laboratory	3.0.0 – 3.6.5
3	Joints, Reinforcement; Embedments and Waterstops; Tolerances; Ordering Concrete	4.0.0 – 8.4.0
4	Formwork; Construct Forms on Grade/Laboratory	3.0.0
5	Module Examination Performance Profile Examination	

Optional References for Advanced Study

This module is intended to present thorough resources for task training. The following reference works are suggested for both instructors and motivated trainees interested in further study. These are optional materials for continued education rather than for task training.

Concrete Construction and Estimating, Craig Avery, Craftsman Book Company, Carlsbad, CA, 1991.

Concrete Manual, International Conference of Building Officials, Whittier, CA, 1992.

Concrete Parking Lots: Eight Steps to Quality Construction, Videotape, Portland Cement Association, Skokie, IL.

Concrete Technology, Third Edition, George R. White, Delmar Publishers, Inc., Albany, NY, 1991.

Design and Control of Concrete Mixtures, Thirteenth Edition, Portland Cement Association, Skokie, IL, 1988.

Finishing Concrete Flatwork, Videotape 32:00 minutes, Portland Cement Association, Skokie, IL, 1991.

Guide for Concrete Floor and Slab Construction, ACI 302.1R-96, American Concrete Institute, Farmington Hills, MI, 1997.

The Contractor's Guide to Quality Construction, American Concrete Institute, Farmington Hills, MI, 1997.

COURSE MAP

This course map shows all of the modules in the first level of the Concrete Finishing curricula. The suggested training order begins at the bottom and proceeds up. Skill levels increase as a trainee advances on the course map. The training order may be adjusted by the local Training Program Sponsor.

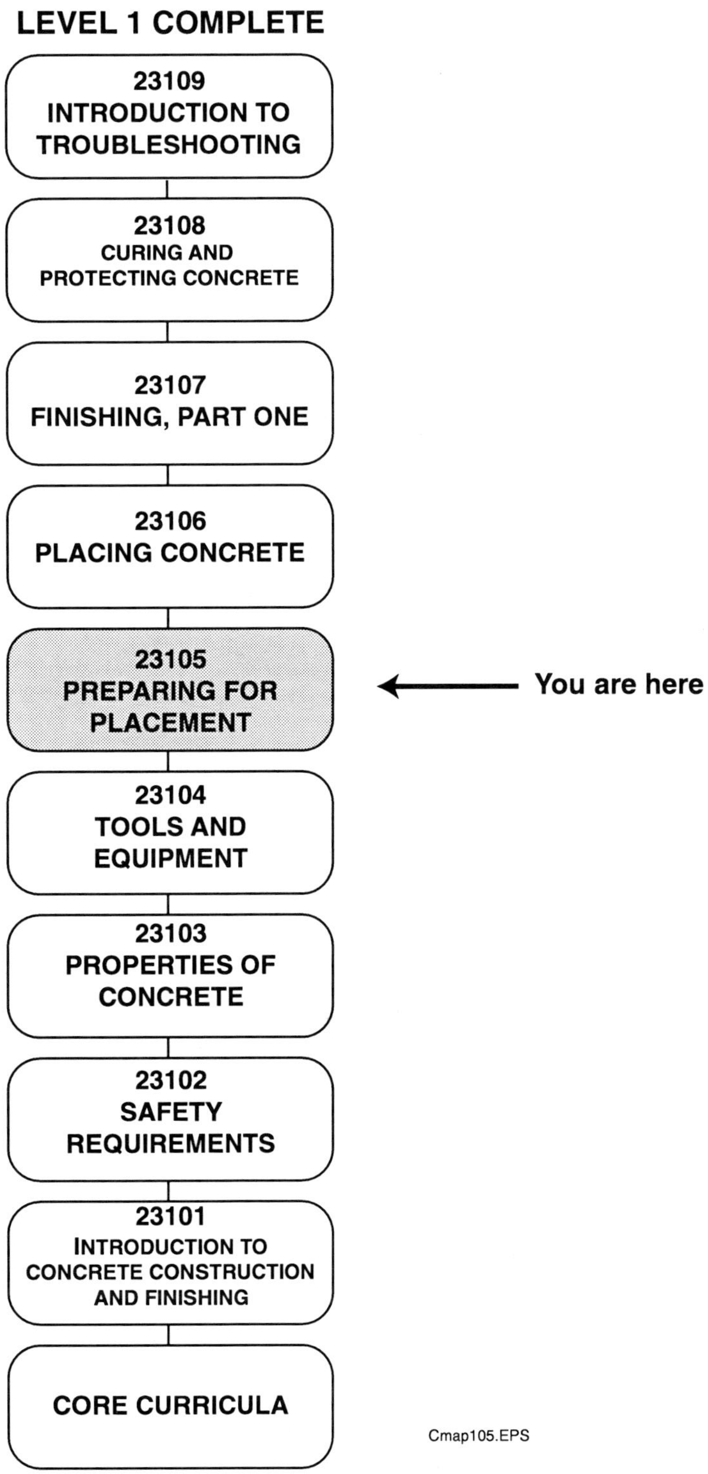

Cmap105.EPS

SESSION 1 OF 5 SESSIONS

Ensure that all trainees have brought their Trainee Modules to class.

Equipment and Materials

Trainee Task Module
Overhead projector and screen
Whiteboard/chalkboard
Markers/chalk
Transparencies
Pencils and paper
Engineer's level and rod
Laser level and rod
Shovels
Construction stakes
Come-alongs
Sandbox(es)
Appropriate Personal Protective Equipment

Time Schedule (150 minutes)

10 Overview
40 Classroom
10 Class Break
30 Classroom
10 Class Break
40 Demonstration
10 Summary

Show Transparency 1, Course Objectives.

Show Transparency 2, Performance Profile Tasks.

Overview (10 minutes)

1. Take attendance and perform other administrative tasks.
2. Outline briefly the topics to be covered. Explain to the trainees that they must pass the Module Examination with a grade of at least 70 percent. Also, explain that they must pass the Performance Profile Examination by performing each task to the level of acceptance criteria.
3. Provide a brief explanation of the acceptance criteria that will be used in the Performance Profile Examination.

Use whiteboard to explain trade terms.

Show Transparency 3, Soft Spots and Hard Spots (Figure 2, Trainee Module).

Classroom (40 minutes)

1. Review the trade terms introduced in this module and explain them.
2. Describe the procedures for subgrade and grade preparation.
   - Proper subgrade conditions
   - Flowable fill

AUDIO/VISUAL

Show Transparency 4, Direct Leveling of Points (Figure 4, Trainee Module).

3. Discuss establishing elevations using:
   - engineer's level
   - laser level

Show Transparency 5, Batter Board (Figure 6, Trainee Module).

4. Explain the set up and use of batter boards.

Show Transparency 6, Installation of Geotextiles and Drainage Pipe (Figure 9, Trainee Module).

5. Explain the use and installation of geotextiles and drainage pipe. Discuss use of these materials in your area.

Class Break (10 minutes)

Classroom (30 minutes)

1. Discuss subgrade and base compacting requirements.
2. Discuss groundwater containment. Explain specific procedures used in your area to reduce groundwater intrusion.

Show Transparency 7, Basic Foundation Types (Figure 10, Trainee Module).

3. Discuss basic types of foundations.
   - Foundation types
   - Constructing foundations

Show Transparency 8, Footing in Trench and Form (Figure 11, Trainee Module).

4. Explain the difference between trench footings and formed footings.

Class Break (10 minutes)

Demonstration (40 minutes)

1. Demonstrate the use of an engineer's level for establishing elevations.
2. Demonstrate the use of a laser level for establishing elevations and setting forms on grade.
3. In a sandbox, demonstrate the use of a shovel and come-along to level and compact subgrade.

Summary (10 minutes)

1. Review the materials covered in this session.
2. Answer trainee questions.
3. Give assignments for the next session.

SESSION 2 OF 5 SESSIONS

Ensure that all trainees have brought their Trainee Modules to class.

Equipment and Materials

Trainee Task Module
Overhead projector and screen
Whiteboard/chalkboard
Markers/chalk
Transparencies
Pencils and paper
Engineer's level and level rods
Laser level and level rods
Sandbox(es)
Shovels
Come-alongs
Appropriate Personal Protective Equipment

Time Schedule (150 minutes)

5 Overview
40 Classroom
10 Class Break
40 Classroom
5 Class Break
45 Laboratory
5 Summary

Overview (5 minutes)

Show Transparency 1, Course Objectives.

Show Transparency 2, Performance Profile Tasks.

1. Take attendance and perform other administrative tasks.
2. Outline briefly the topics to be covered. Explain to the trainees that they must pass the Module Examination with a grade of at least 70 percent. Also, explain that they must pass the Performance Profile Examination by performing each task to the level of acceptance criteria.
3. Provide a brief explanation of the acceptance criteria that will be used in the Performance Profile Examination.

Classroom (40 minutes)

1. Discuss the importance of forms to concrete work.
2. Describe the various types of forms used.
   - Job-built, single-use wall form

Show Transparency 9, Typical Job-Built Wall Form (Figure 12, Trainee Module).

AUDIO/VISUAL

Show Transparency 10, Low Footing Form (Figure 13, Trainee Module).

- Job-built, single-use footing form

Show Transparency 11, Ganged Forms (Figure 14, Trainee Module).

- Prefabricated, multiple-use wall form

Show Transparency 12, Nail-Down Pan (Figure 15, Trainee Module).

- Manufactured pans

Show Transparency 13, Prefabricated Forms for Slabs on Grade (Figure 18, Trainee Module).

3. Discuss prefabricated forms for slabs on grade.
4. Discuss form hardware and release agents.
5. Explain how to square up forms.

Show Transparency 14, Setting Grade for Slab (Figure 19, Trainee Module).

6. Explain how to set forms to line and grade.
7. Describe the procedure used for inspecting forms.

Class Break (10 minutes)

Classroom (40 minutes)

1. Describe the process for laying out and constructing forms for a slab on grade.

Show Transparency 15, Finding the Baseline (Figure 20, Trainee Module).

2. Discuss establishing baseline.

Show Transparency 16, Staking the First Two Corners (Figure 21, Trainee Module).

3. Explain how to establish corners.

Show Transparency 17, Staking for Batter Boards (Figure 22, Trainee Module).

4. Explain how to set batter boards for use in excavation of a foundation or slab.

Show Transparency 18, Dimension Lumber Sizes and Slab Thickness (Figure 23, Trainee Module).

Show Transparency 19, Brace Stakes (Figure 24, Trainee Module).

Make sure levels are calibrated before using.

5. Describe placing and securing form boards.

Class Break (5 minutes)

Laboratory (45 minutes)

1. Supervise and monitor trainees as they practice using an engineer's level and/or laser level and level rod.
2. Supervise and monitor trainees as they practice using a shovel and come-along to compact a subgrade.
3. Ensure that all safety procedures are followed.

Summary (5 minutes)

1. Review the materials covered in this session.
2. Answer trainee questions.
3. Give assignments for the next session.

SESSION 3 OF 5 SESSIONS

Ensure that all trainees have brought their Trainee Modules to class.

Equipment and Materials

Trainee Task Module
Overhead projector and screen
Whiteboard/chalkboard
Markers/chalk
Transparencies
Hand level
Engineer's level or laser level
Level rod
Measuring tape
Site plan
Hammers
Nails
Construction stakes
Batter boards
Grease crayon or marking pen
Reinforcing steel
Waterstop
String (mason's line) and cutter
Appropriate Personal Protective Equipment

Time Schedule (150 minutes)

5 Overview
60 Classroom
10 Class Break
30 Demonstration
5 Class Break
30 Laboratory
10 Summary

Show Transparency 1, Course Objectives.

Show Transparency 2, Performance Profile Tasks.

Overview (5 minutes)

1. Take attendance and perform other administrative tasks.
2. Outline briefly the topics to be covered. Explain to the trainees that they must pass the Module Examination with a grade of at least 70 percent. Also, explain that they must pass the Performance Profile Examination by performing each task to the level of acceptance criteria.
3. Provide a brief explanation of the acceptance criteria that will be used in the Performance Profile Examination.

AUDIO/VISUAL

Classroom (60 minutes)

Show Transparency 20, Outside Locations of Joints (Figure 26, Trainee Module).

1. Describe the different types of joints that are made in concrete. Explain the need for each type.

Show Transparency 21, Sawcut and Preformed Contraction Joints (Figure 27, Trainee Module).

- Contraction joints

Show Transparency 22, Joints at Re-entrant Corners (Figure 28, Trainee Module).

- Contraction joints placed at re-entrant corners

Show Transparency 23, Isolation Joints Around Columns (Figure 30, Trainee Module).

- Isolation joints

Show Transparency 24, Types of Construction Joints (Figure 31, Trainee Module).

- Construction joints

Show Transparency 25, Standard Deformed Reinforcing Bars (Figure 34, Trainee Module).

2. Explain the significance of reinforcing steel in concrete construction. Describe the different types of steel reinforcement.
   - Rod
   - Wire mesh
3. Review requirements for positioning reinforcement in forms.
4. Describe the placement of embedments and waterstops.
   - Embedded items
   - Anchor bolts
   - Waterstops

Show Transparency 26, Waterstops (Figure 39, Trainee Module).

Show Transparency 27, ACI Standard Specifications for Tolerances (Example) (Figure 40, Trainee Module).

5. Review tolerance requirements. Discuss specifications and other requirements used in your area.

6. Explain the requirements for ordering concrete from a batch plant. Review the processes for:
    - Estimating quantities
    - Mix design
    - Establishing site access
    - Delivery schedule
7. Describe all the activities that should go into inspection of the site before the concrete is delivered.
    - Subgrade
    - Formwork
    - Reinforcement

Class Break (10 minutes)

Demonstration (30 minutes)

1. Demonstrate how to set up line and grade for a slab using batter boards and a string line. Show how to set corners and determine elevations for a footing.

Class Break (5 minutes)

Laboratory (30 minutes)

1. Have trainees set up batter boards for a small slab as shown on a site plan.
2. Monitor work and make any corrections as necessary.
3. Ensure that all Appropriate Personal Protective Equipment is used correctly.

Summary (10 minutes)

1. Review the material covered in this session.
2. Give assignments for the next session.

Ensure that all trainees have brought their Trainee Modules to class.

Make sure there are sufficient supplies of equipment and material for the laboratory.

Show Transparency 1, Course Objectives.

Show Transparency 2, Performance Profile Tasks.

SESSION 4 OF 5 SESSIONS

Equipment and Materials

Trainee Task Module
Overhead projector and screen
Whiteboard/chalkboard
Markers/chalk
Transparencies
Sandbox(es)
Engineer's level or laser level
Level rods
Hammers
Nails
Construction stakes
Dimensional lumber (2 × 6)
Grease crayon or marking pen
String (mason's line) and cutter
Appropriate Personal Protective Equipment

Time Schedule (150 minutes)

10 Overview
30 Demonstration
10 Class Break
90 Laboratory
10 Summary

Overview (10 minutes)

1. Take attendance and perform other administrative tasks.
2. Outline briefly the topics to be covered. Explain to the trainees that they must pass the Module Examination with a grade of at least 70 percent. Also, explain that they must pass the Performance Profile Examination by performing each task to the level of acceptance criteria.
3. Provide a brief explanation of the acceptance criteria that will be used in the Performance Profile Examination.
4. Review the steps in the demonstration from the previous class.
5. Answer any questions about the previous demonstration.

Demonstration (30 minutes)

1. Demonstrate how to locate and construct a form for a slab on grade using a level and string line. Construct a form using 2 × 6 lumber and construction stakes.

Class Break (10 minutes)

Laboratory (90 minutes)

1. Divide trainees into groups of two to four.
2. Assign trainees an area to practice construction of a form for a slab on grade.
3. Observe and monitor trainee performance for the following activities:
    - Locate site and grade for form
    - Construct form
    - Use shovel and come-along to level subgrade inside form

4. Ensure that all Appropriate Personal Protective Equipment is used correctly.
5. Have trainees who are finished with the laboratory work on the Review/Practice Questions at the back of the Trainee Task Module.

Summary (10 minutes)

1. Review the procedures and activities used during the laboratory.
2. Explain that the next session will be the last session and includes the Module Examination and Performance Profile Examination.

SESSION 5 OF 5 SESSIONS

Ensure that all trainees have brought their Trainee Modules to class.

Equipment and Materials

Trainee Task Module
Overhead projector and screen
Whiteboard/chalkboard
Markers/chalk
Transparencies
Module Examination
Performance Profile Sheets
Sandbox(es)
Engineer's level or laser level
Level rods
Hammers
Shovels
Come-alongs
Nails
Construction stakes
Dimensional lumber (2 × 4 and 2 × 6)
Grease crayon or marking pen
String (mason's line) and cutter
Appropriate Personal Protective Equipment

Time Schedule (150 minutes)

20 Overview
30 Module Examination
10 Class Break
90 Performance Profile Examination

Overview (20 minutes)

Show Transparency 1, Course Objectives.

Show Transparency 2, Performance Profile Tasks.

1. Take attendance and perform other administrative tasks.
2. Outline briefly the topics to be covered. Explain to the trainees that they must pass the Module Examination with a grade of at least 70 percent. Also, explain that they must pass the Performance Profile Examination by performing each task to the level of acceptance criteria.
3. Provide a brief explanation of the acceptance criteria that will be used in the Performance Profile Examination

Module Examination (30 minutes)

1. Hand out Module Examinations to all trainees. Tell them they have 30 minutes to complete the exam.
2. Collect and grade the examinations.

Class Break (10 minutes)

Performance Profile Examination (90 minutes)

1. Have trainees perform the following performance objectives to the satisfaction of the instructor:
   - Set a batter board
   - Locate and level a form
   - Build a slab form to grade with dimension lumber
   - Compact subgrade with a shovel and come-along
2. Fill out Performance Profile Sheets for each trainee.

MODULE EXAMINATIONS AND ANSWER KEYS

The masters for Module Examinations and Answer Keys are found in the booklet entitled "Concrete Finishing Level One Module Examinations" included in this Instructor's Guide package. You may also reproduce the tests using the Windows-compatible software program which may be purchased separately from Prentice Hall. If you do not have the reproducible masters in your Instructor's Guide package, or would like to order the test software, please contact Prentice Hall Publishers at 1-800-922-0579.

Be sure to keep answer keys and tests in a safe area.

OVERHEAD TRANSPARENCIES

Masters for overhead transparencies are found on the following pages. Transparencies can be made by using clear acetate sheets in a photocopy machine. This can be done at any commercial copy shop, or by following this method on your office machine:

1. Use standard 8½ × 11 clear acetate sheets, which can be obtained at office supply stores, stationery stores, or commercial copying centers.
2. Check the photocopier's manual to make sure it will take the thicker acetate sheets.
3. Remove the paper supply in the photocopying machine and replace it with a supply of acetate sheets.
4. Make photocopies from the masters on the following pages in the usual way.
   The material will be reproduced on the clear acetate sheets for use on any overhead projector.

TRANSPARENCY 1

TASK MODULE 23105, PREPARING FOR PLACEMENT

COURSE OBJECTIVES

Upon completion of this module, the trainee will be able to:

1. **Describe basic site layout using levels and measuring tools.**
2. **Properly locate, grade, and build forms for horizontal placement.**
3. **Perform compaction activities on subgrades.**
4. **Describe various joints and where to locate them.**
5. **Describe various reinforcements and how to place them.**
6. **Describe information needed when ordering concrete.**

TRANSPARENCY 2

TASK MODULE 23105, PREPARING FOR PLACEMENT

PERFORMANCE PROFILE TASKS

1. Set a batter board.
2. Locate and level a form.
3. Build a slab form to grade with dimension lumber.
4. Compact subgrade with a shovel and come-along.

TRANSPARENCY 3

FIGURE 2

TASK MODULE 23105, PREPARING FOR PLACEMENT

SOFT SPOTS AND HARD SPOTS

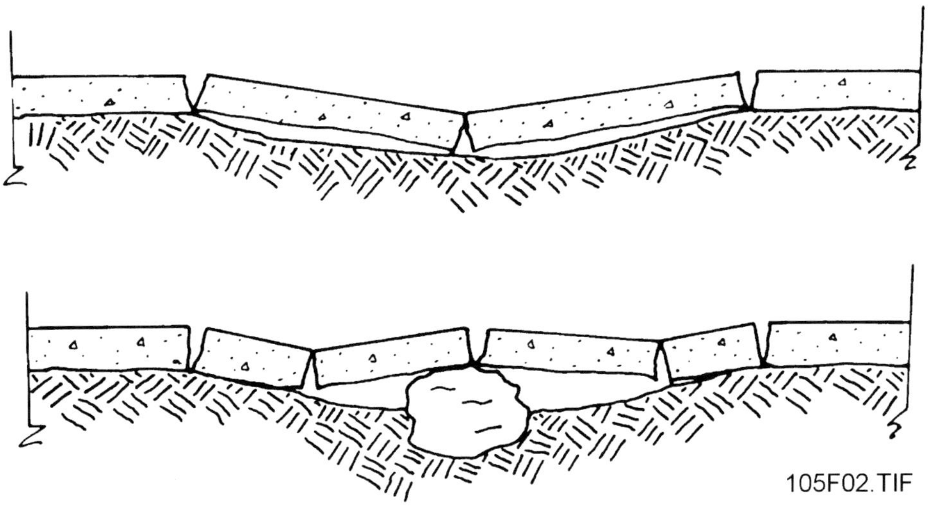

105F02.TIF

TRANSPARENCY 4

FIGURE 4

TASK MODULE 23105, PREPARING FOR PLACEMENT

DIRECT LEVELING OF POINTS

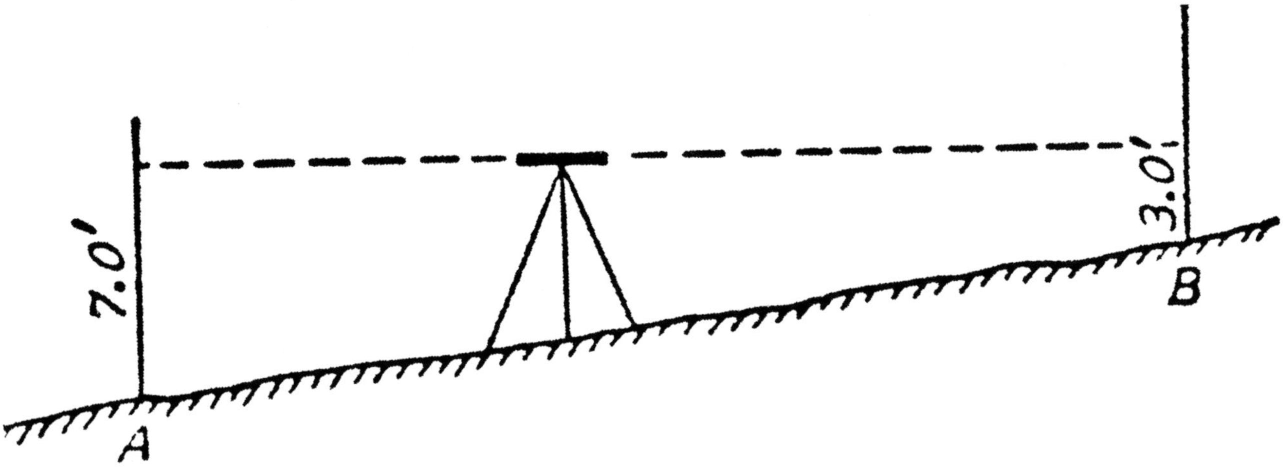

105F04.TIF

TRANSPARENCY 5

FIGURE 6

TASK MODULE 23105, PREPARING FOR PLACEMENT

BATTER BOARD

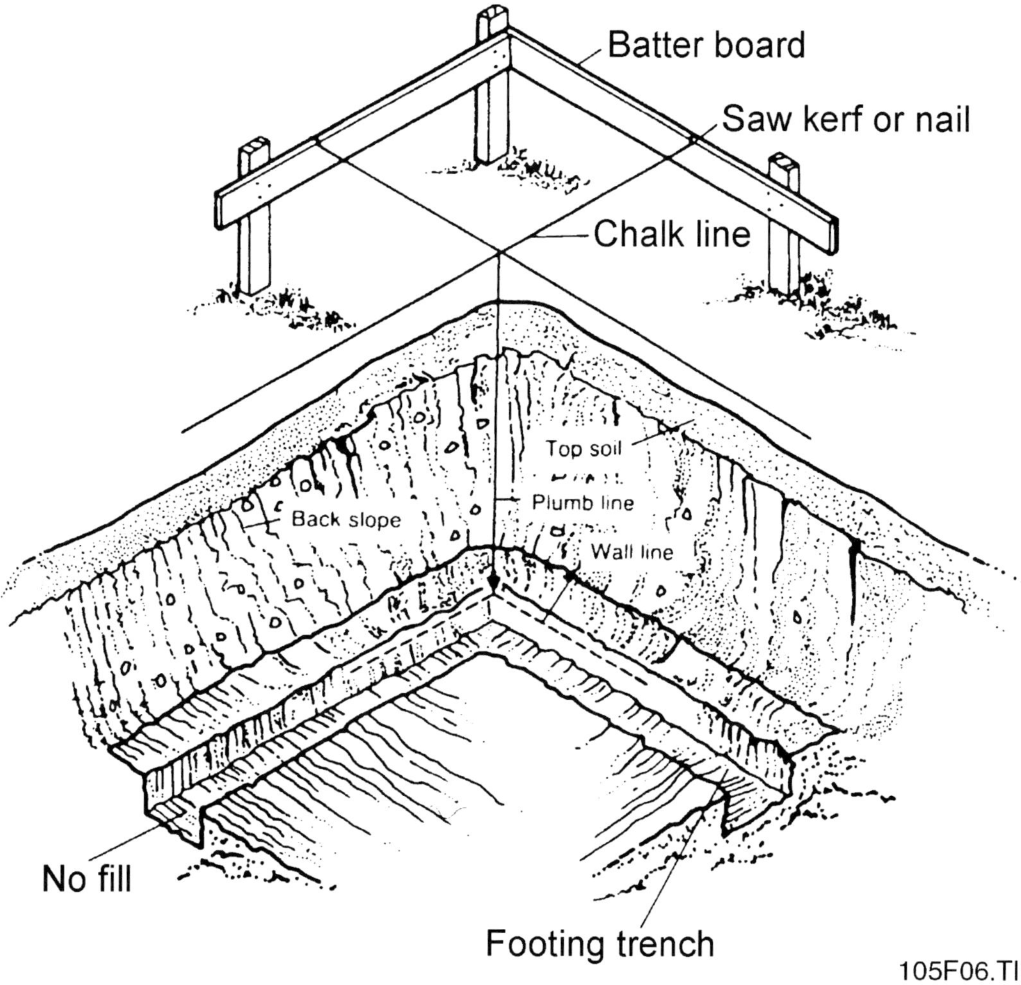

105F06.TIF

TRANSPARENCY 6

FIGURE 9

TASK MODULE 23105, PREPARING FOR PLACEMENT

INSTALLATION OF GEOTEXTILES AND DRAINAGE PIPE

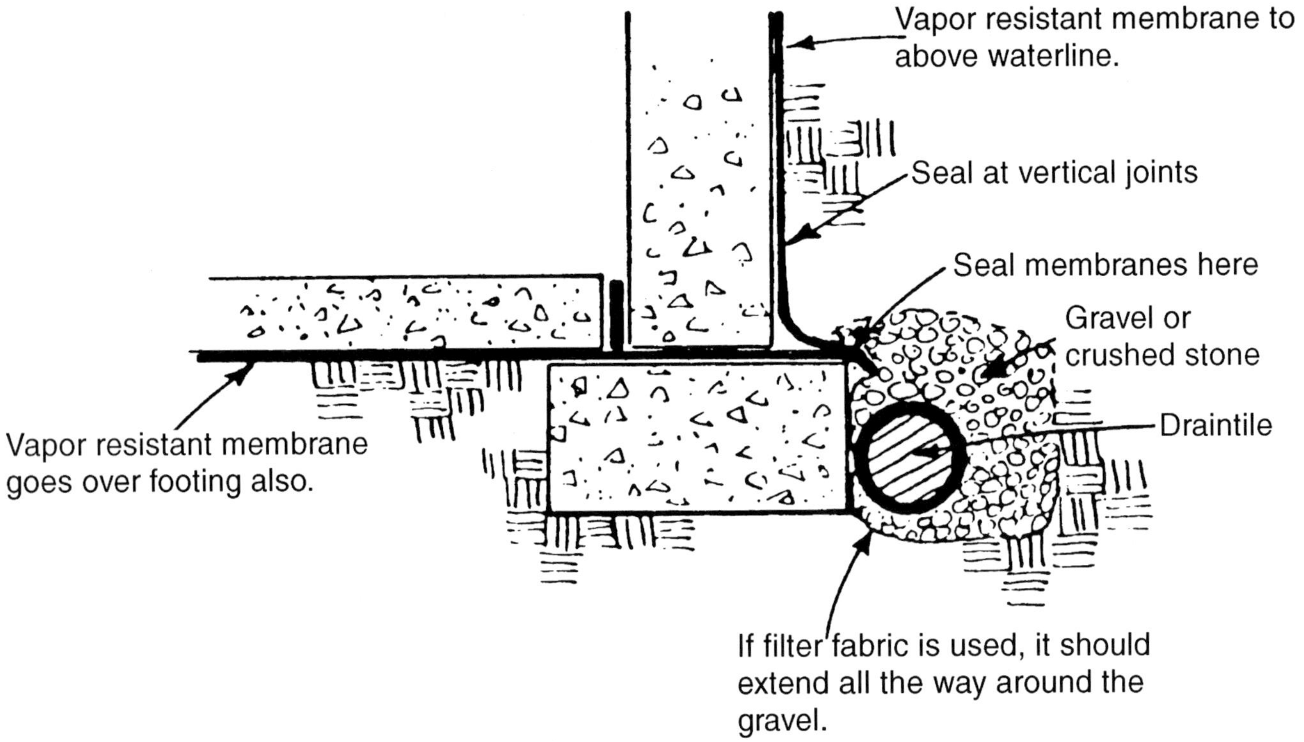

105F09.TIF

FIGURE 10

TASK MODULE 23105, PREPARING FOR PLACEMENT

BASIC FOUNDATION TYPES

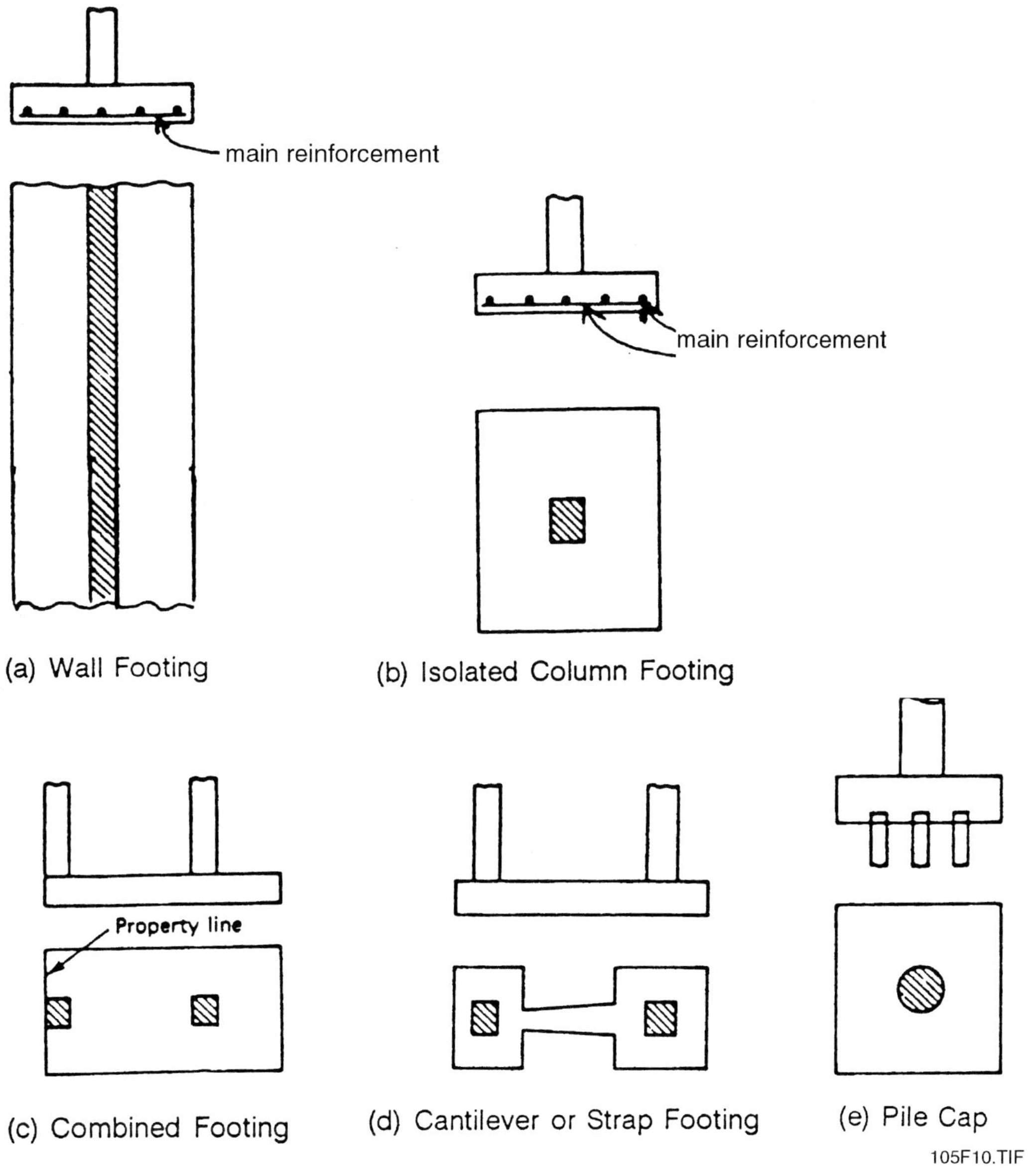

TRANSPARENCY 8

FIGURE 11

TASK MODULE 23105, PREPARING FOR PLACEMENT

FOOTING IN TRENCH AND FORM

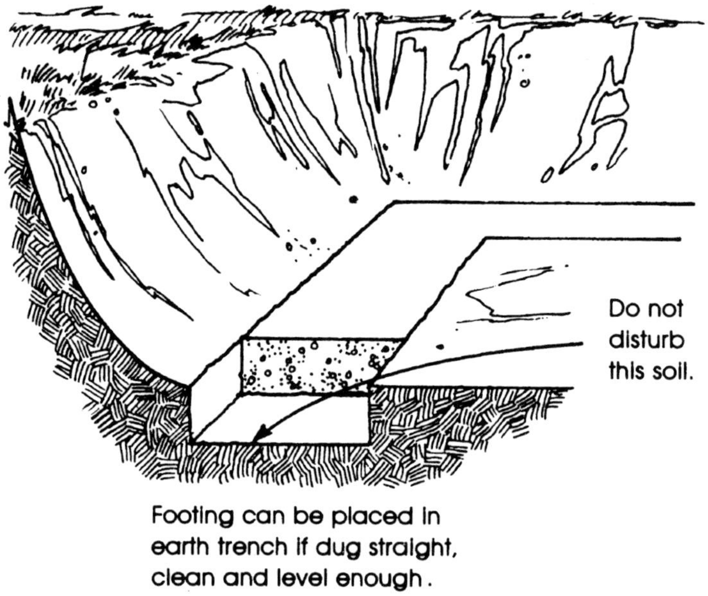

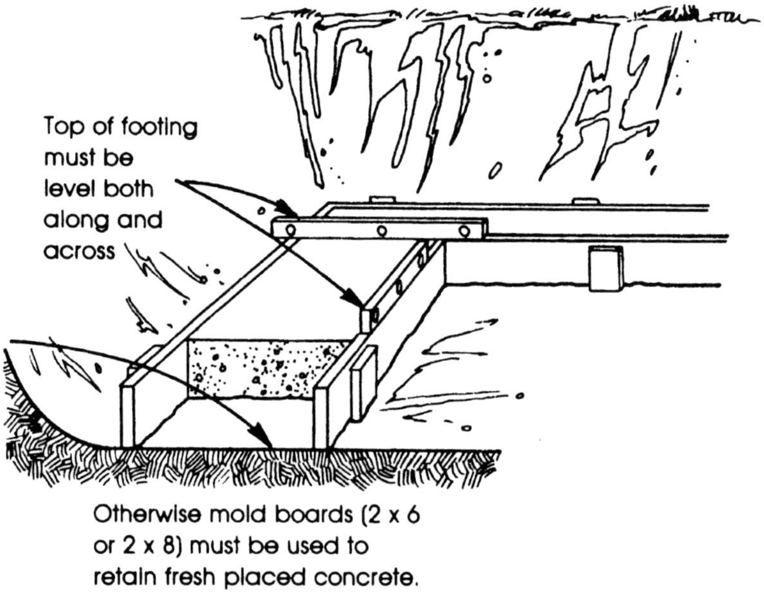

105F11.TIF

TRANSPARENCY 9

FIGURE 12

TASK MODULE 23105, PREPARING FOR PLACEMENT

TYPICAL JOB-BUILT WALL FORM

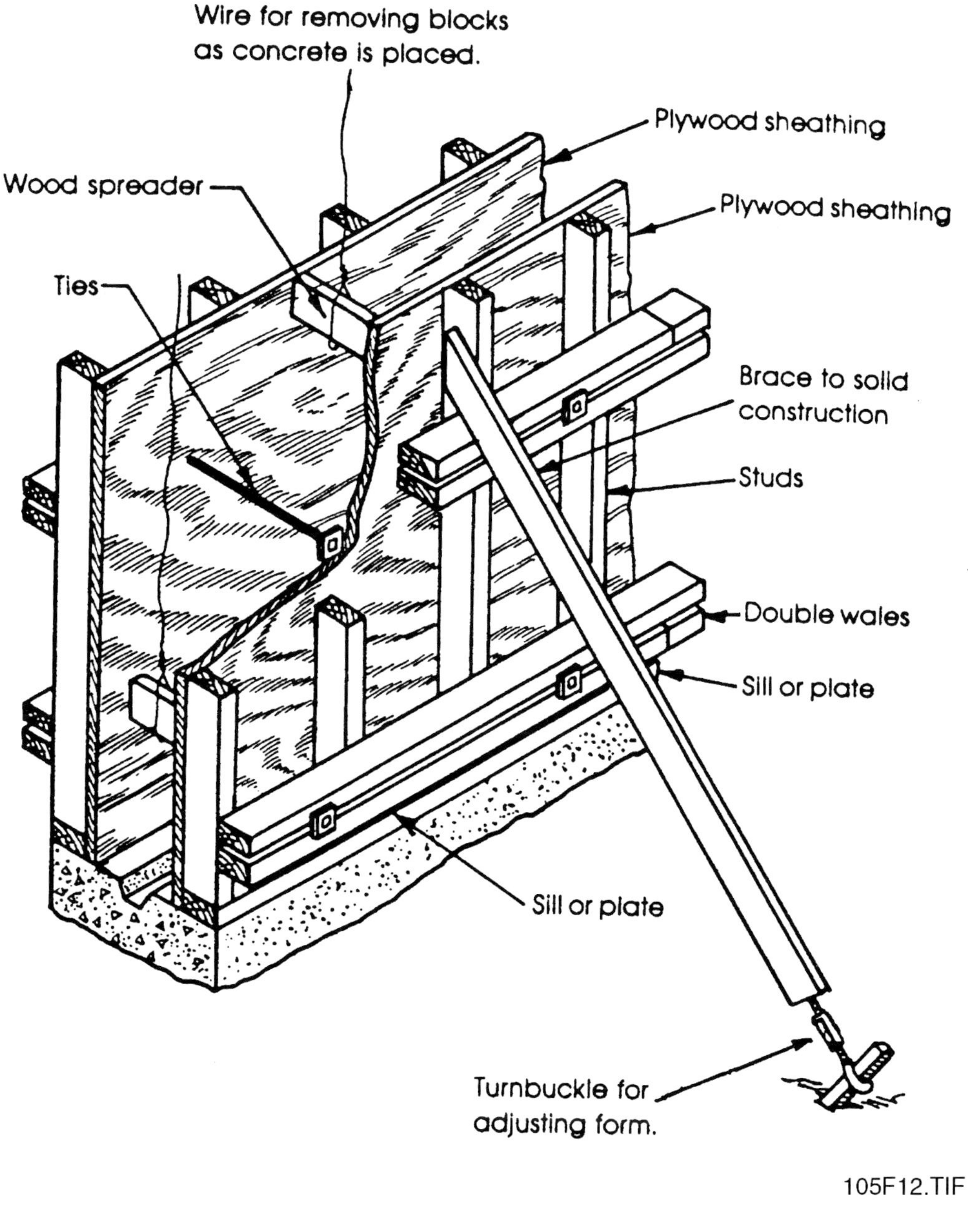

105F12.TIF

TRANSPARENCY 10

FIGURE 13

TASK MODULE 23105, PREPARING FOR PLACEMENT

LOW FOOTING FORM

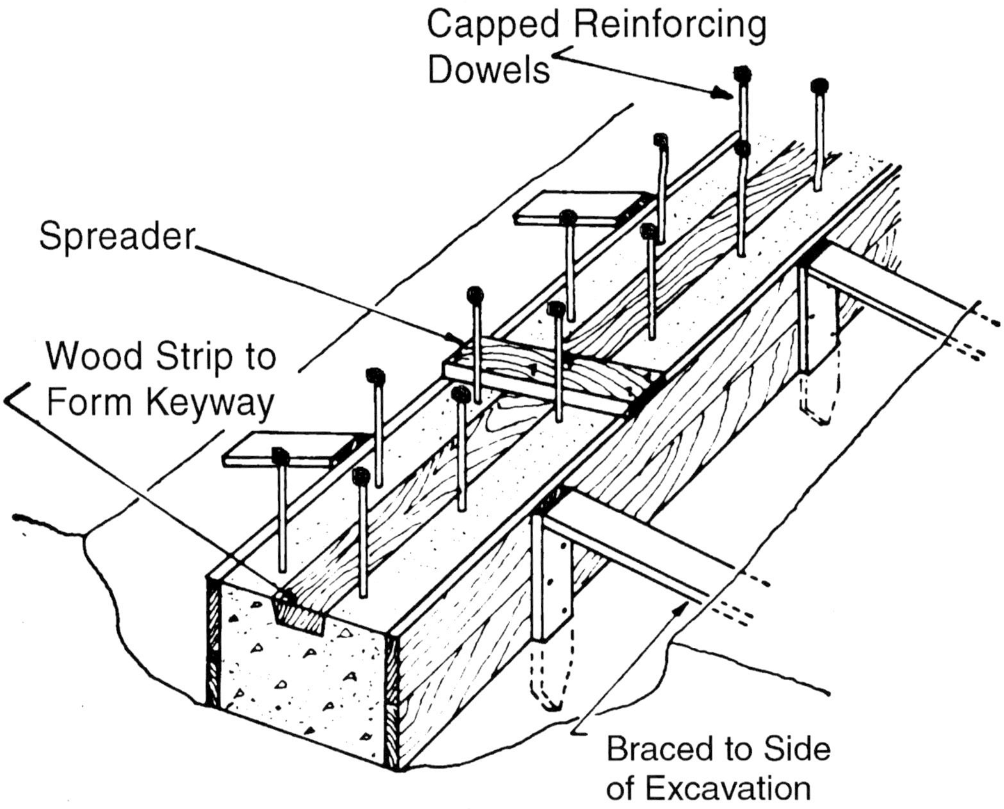

105F13.TIF

TRANSPARENCY 11

FIGURE 14

TASK MODULE 23105, PREPARING FOR PLACEMENT

GANGED FORMS

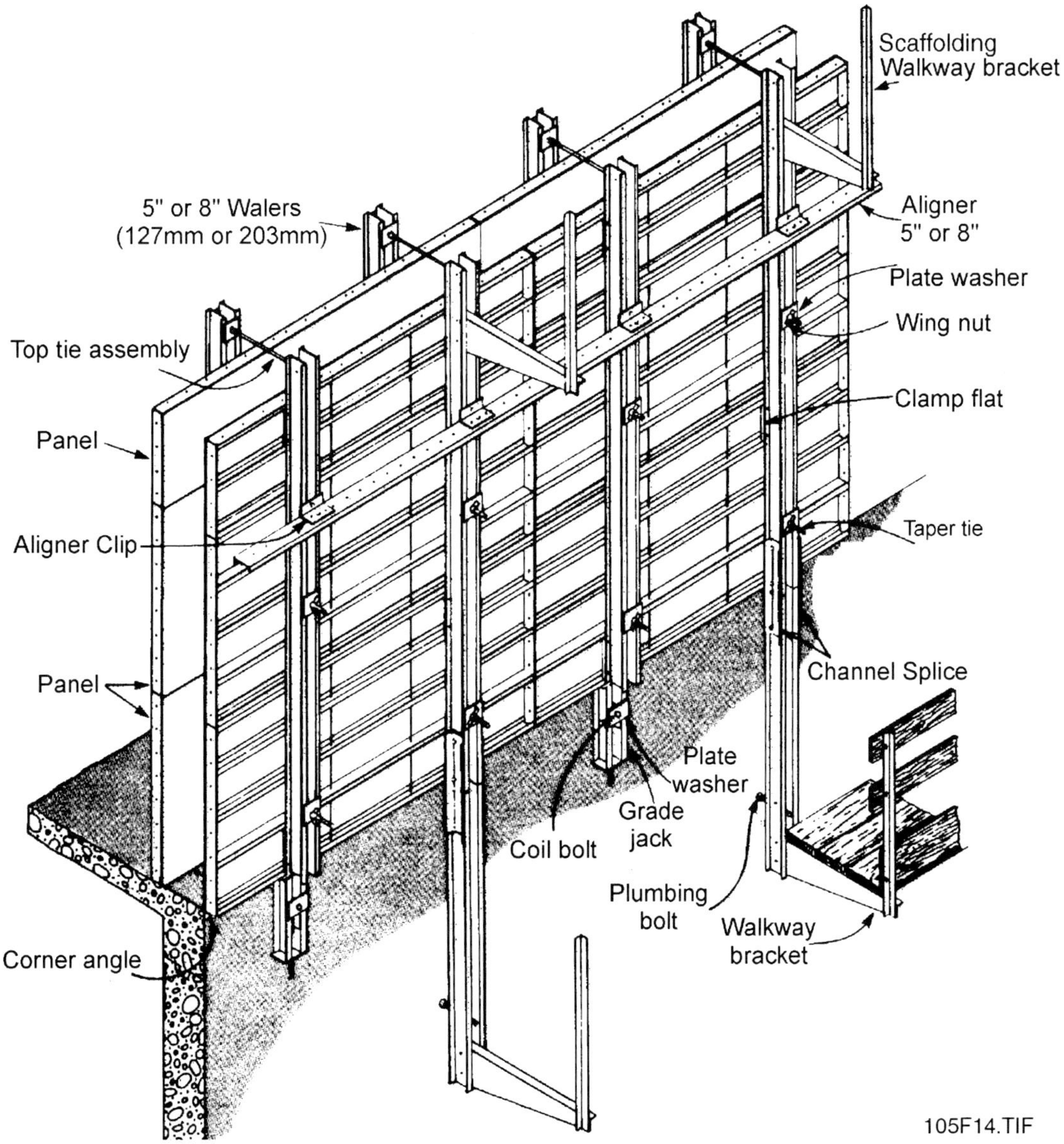

TRANSPARENCY 12

FIGURE 15

TASK MODULE 23105, PREPARING FOR PLACEMENT

NAIL-DOWN PAN

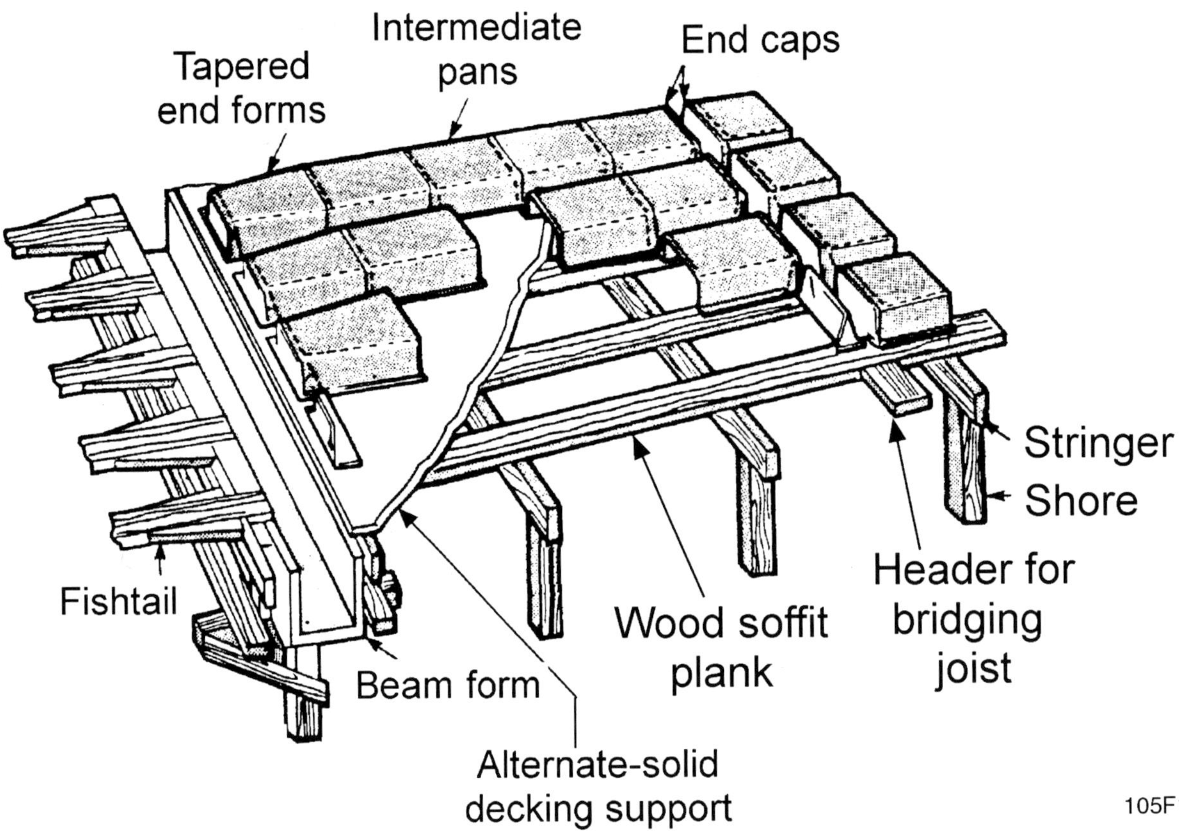

105F15.TIF

TRANSPARENCY 13

FIGURE 18

TASK MODULE 23105, PREPARING FOR PLACEMENT

PREFABRICATED FORMS FOR SLABS ON GRADE

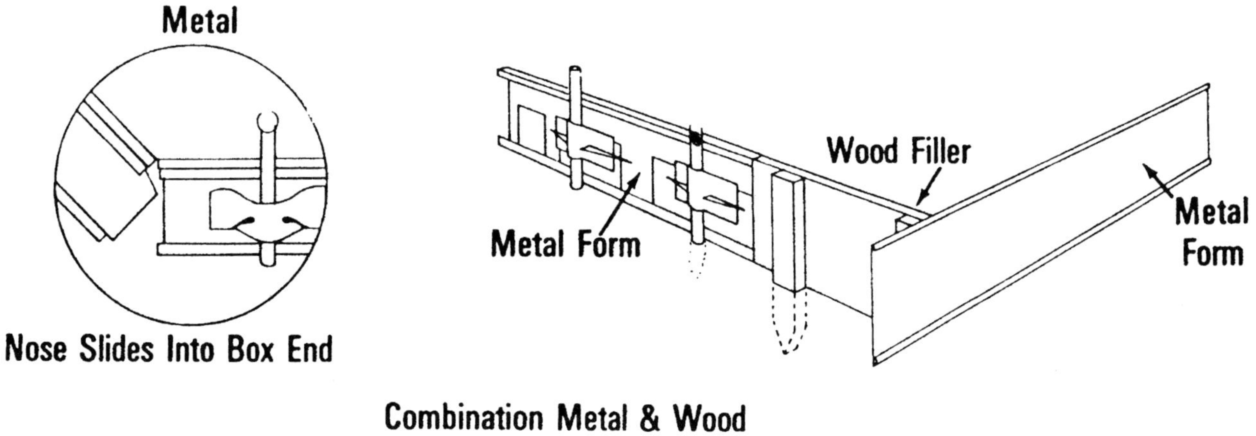

105F18.TIF

TRANSPARENCY 14

FIGURE 19

TASK MODULE 23105, PREPARING FOR PLACEMENT

SETTING GRADE FOR SLAB

105F19.TIF

TRANSPARENCY 15

FIGURE 20

TASK MODULE 23105, PREPARING FOR PLACEMENT

FINDING THE BASELINE

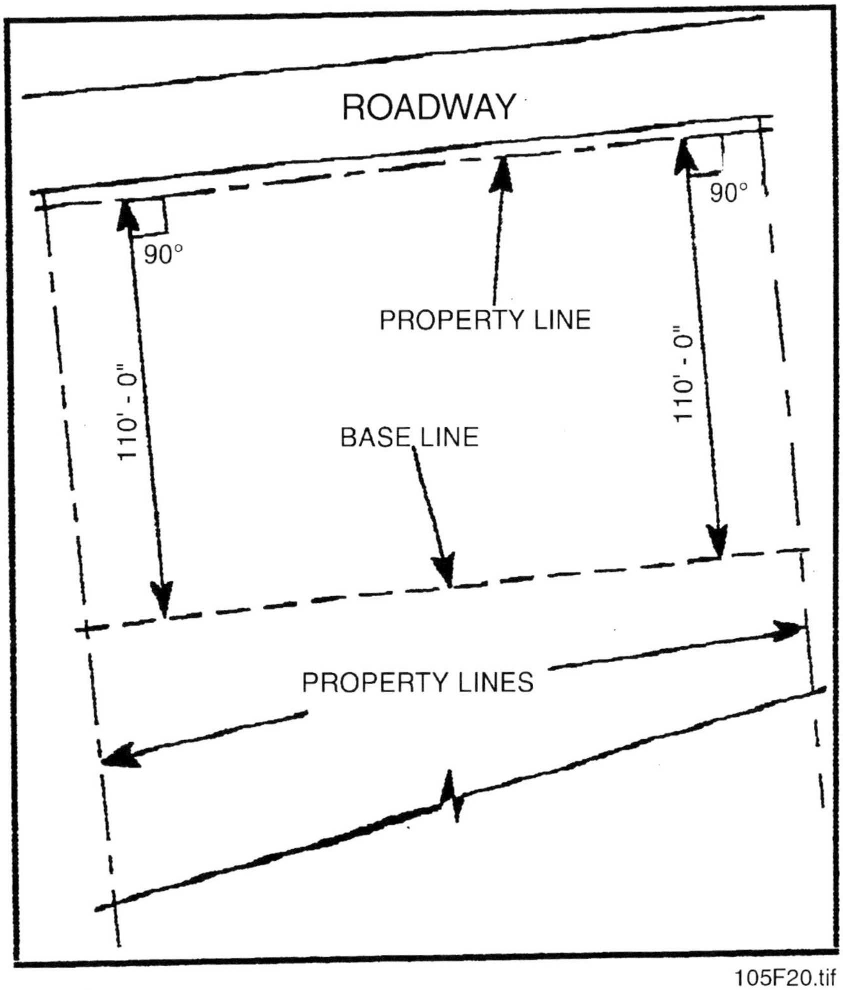

105F20.tif

TRANSPARENCY 16

FIGURE 21

TASK MODULE 23105, PREPARING FOR PLACEMENT

STAKING THE FIRST TWO CORNERS

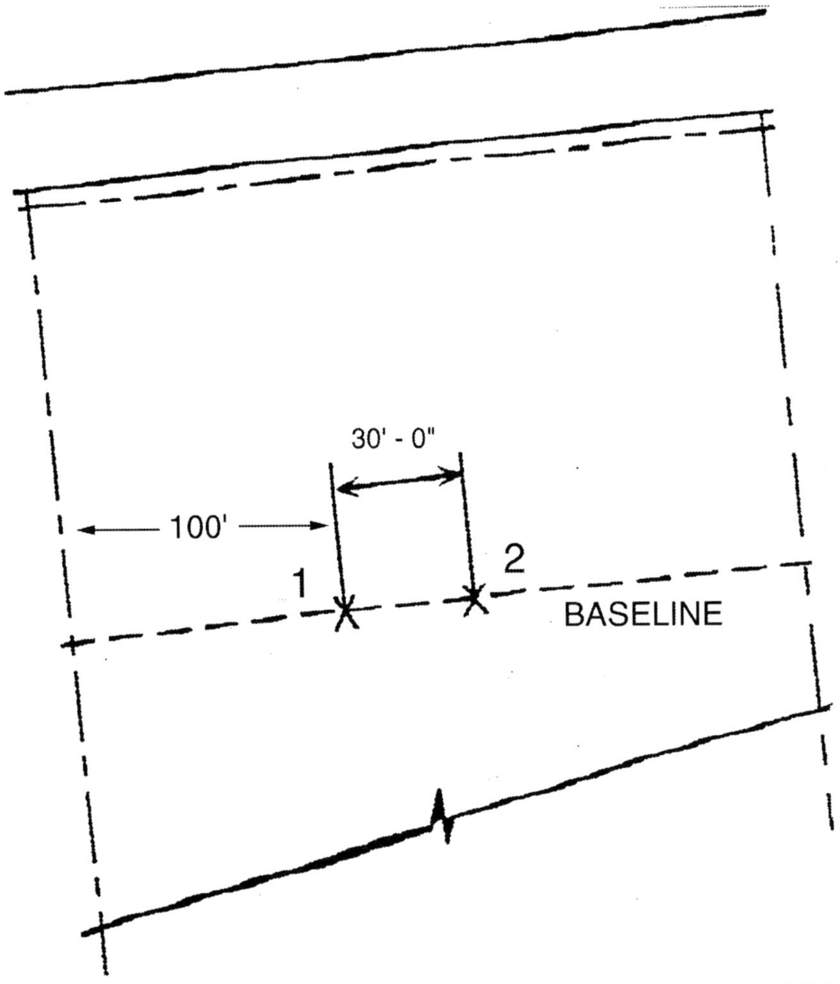

105F21.tif

TRANSPARENCY 17

FIGURE 22

TASK MODULE 23105, PREPARING FOR PLACEMENT

STAKING FOR BATTER BOARDS

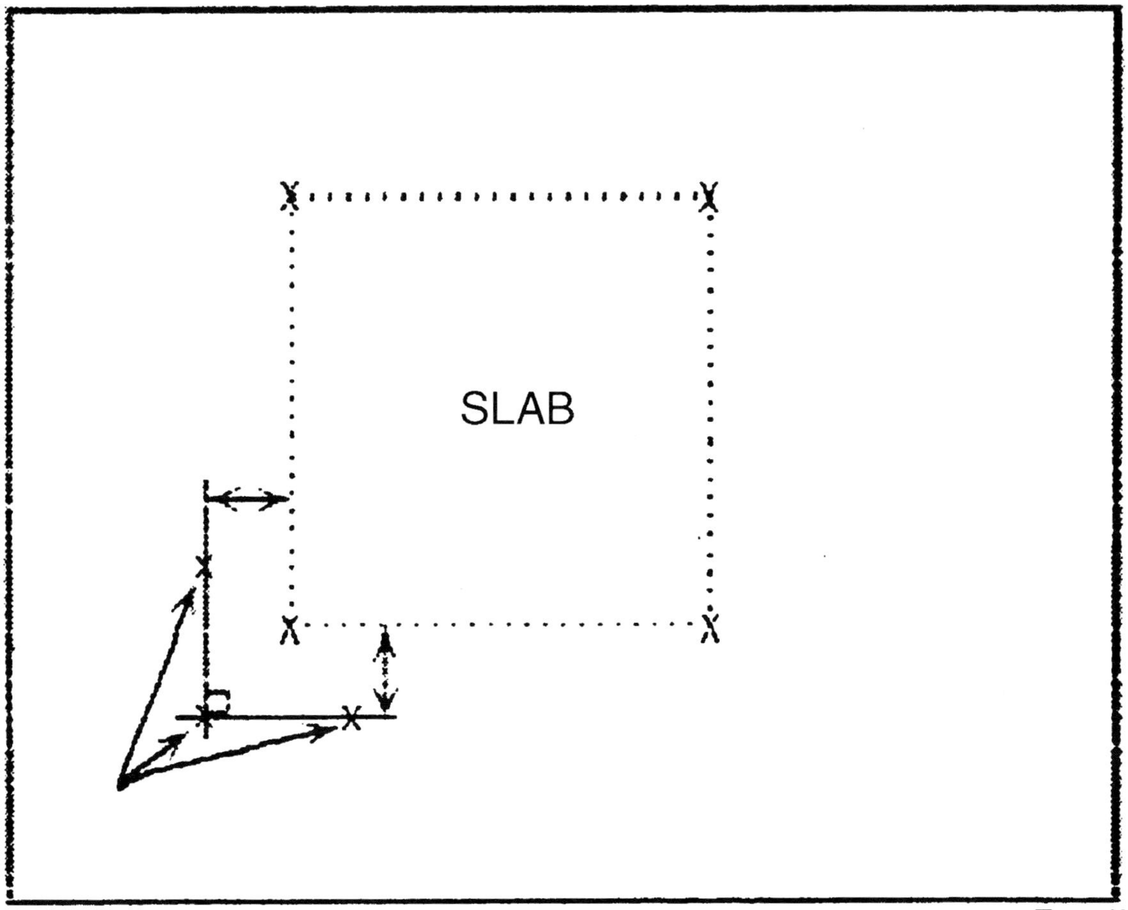

105F22.tif

TRANSPARENCY 18

FIGURE 23

TASK MODULE 23105, PREPARING FOR PLACEMENT

DIMENSION LUMBER SIZES AND SLAB THICKNESS

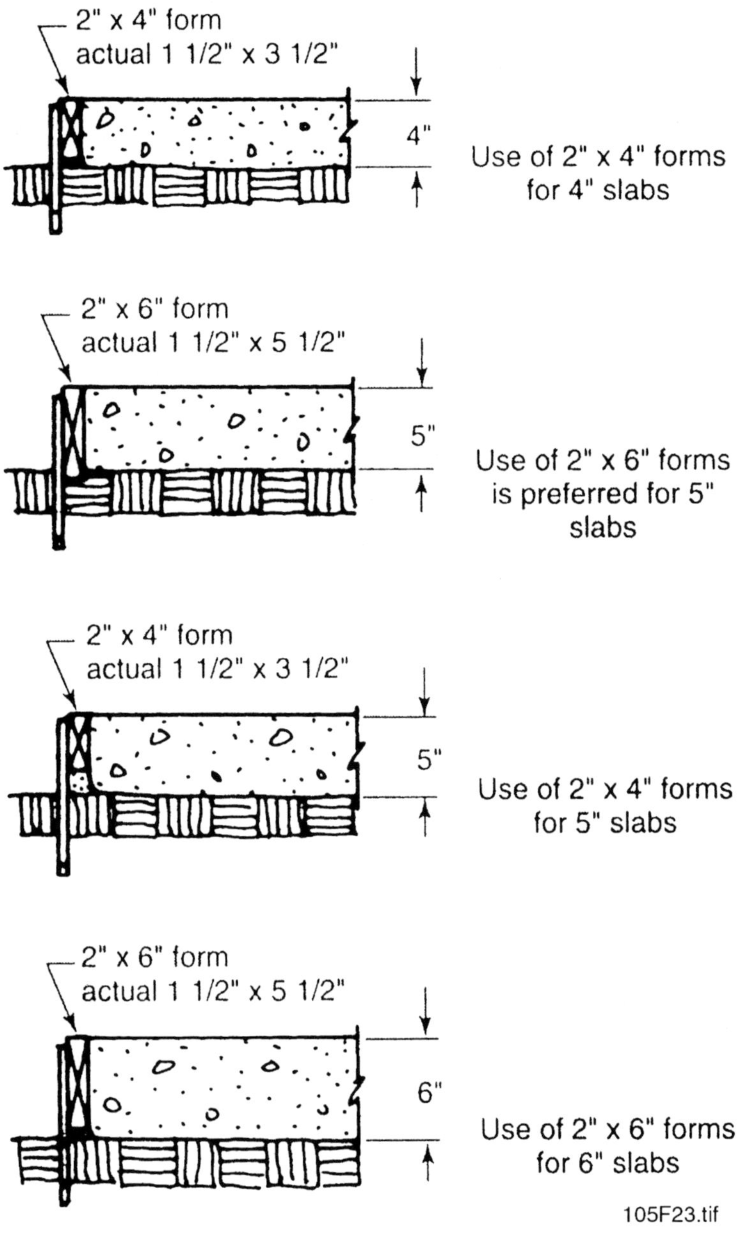

TRANSPARENCY 19

FIGURE 24

TASK MODULE 23105, PREPARING FOR PLACEMENT

BRACE STAKES

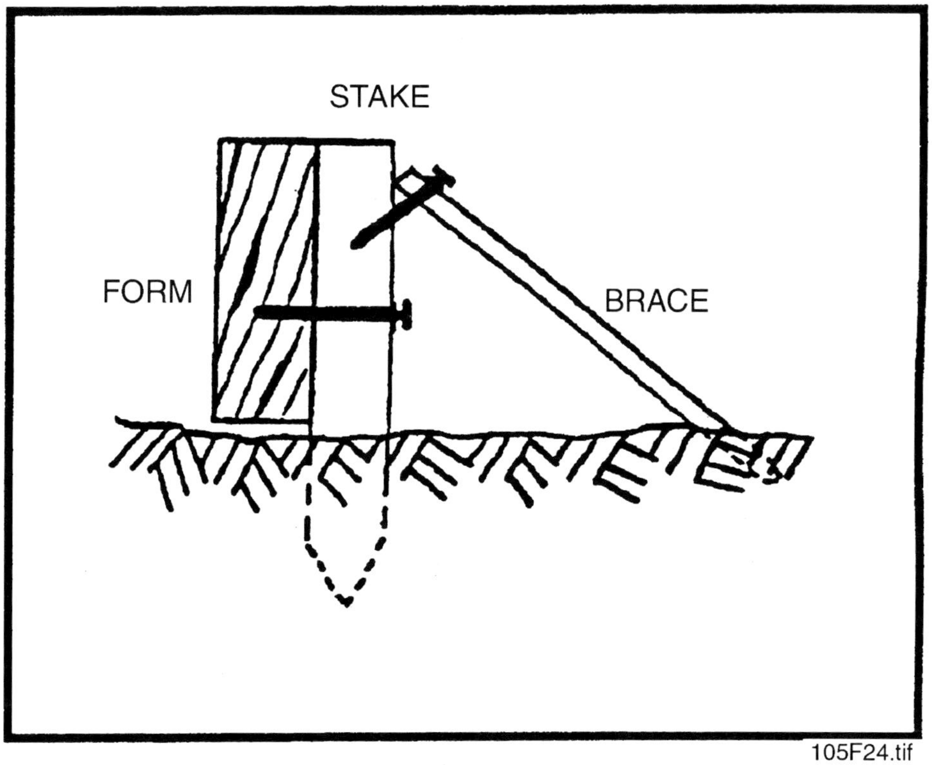

105F24.tif

TRANSPARENCY 20

FIGURE 26

TASK MODULE 23105, PREPARING FOR PLACEMENT

OUTSIDE LOCATIONS OF JOINTS

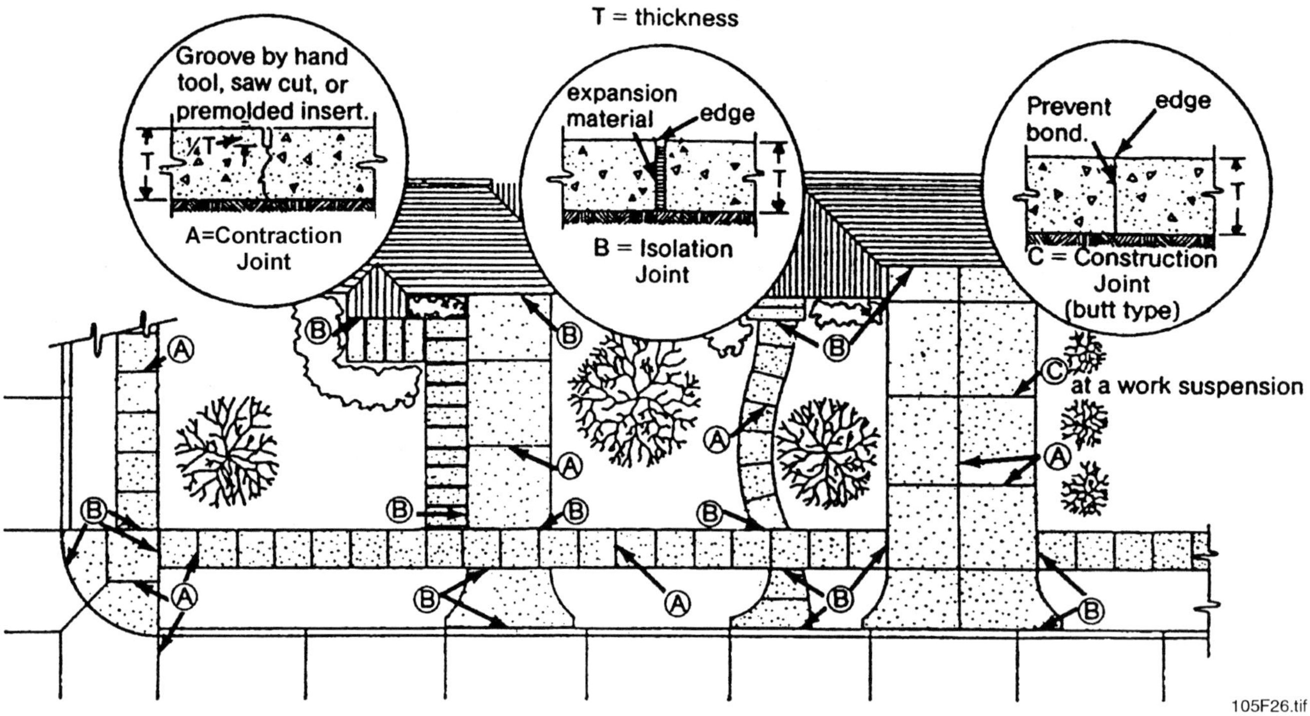

TRANSPARENCY 21

FIGURE 27

TASK MODULE 23105, PREPARING FOR PLACEMENT

SAWCUT AND PREFORMED CONTRACTION JOINTS

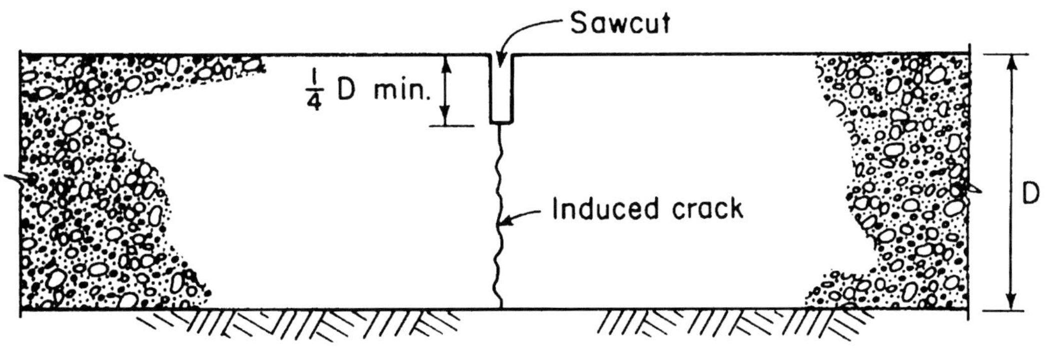

Sawed contraction joint

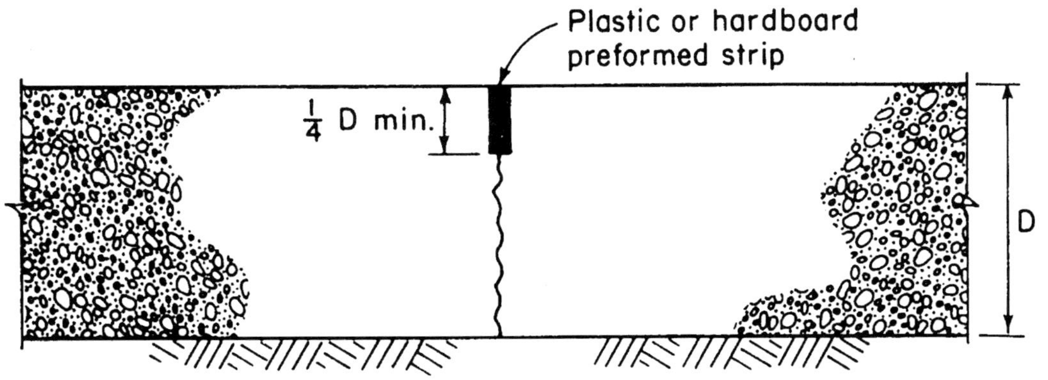

Premolded insert contraction joint

105F27.tif

TRANSPARENCY 22

FIGURE 28

TASK MODULE 23105, PREPARING FOR PLACEMENT

JOINTS AT RE-ENTRANT CORNERS

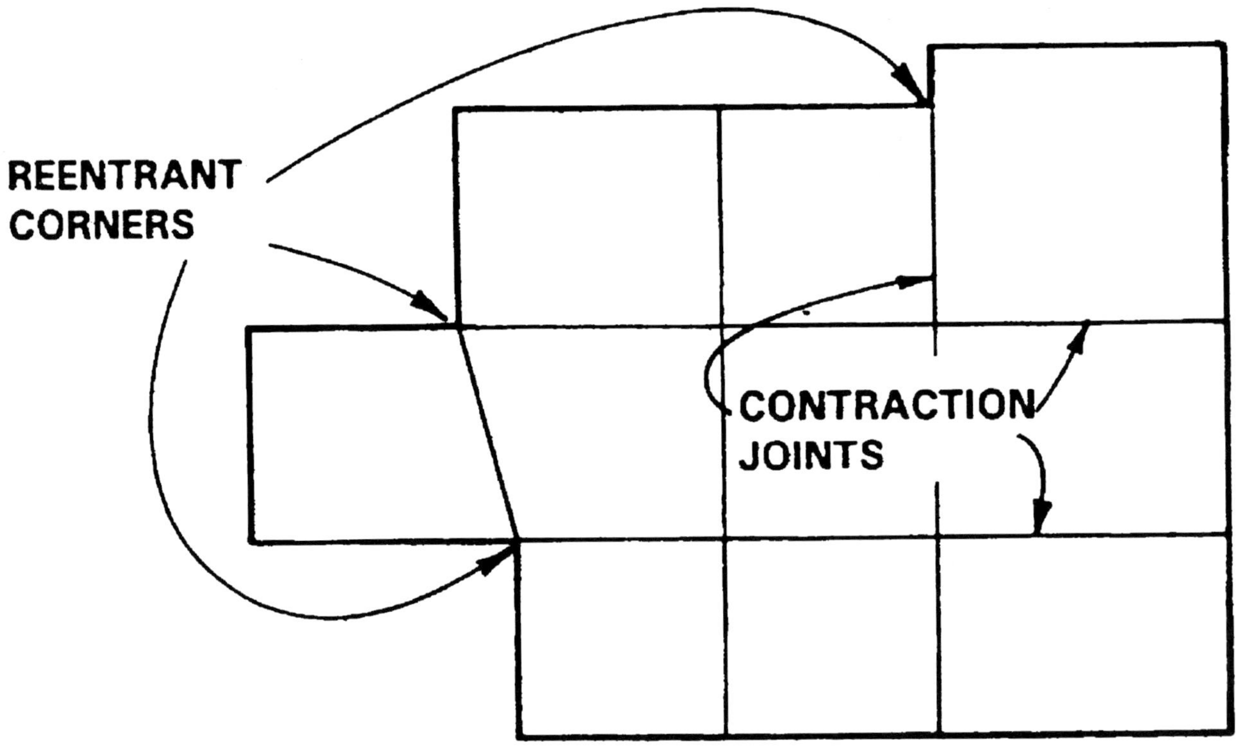

Plan of Slab on Grade

105F28.tif

TRANSPARENCY 23

FIGURE 30

TASK MODULE 23105, PREPARING FOR PLACEMENT

ISOLATION JOINTS AROUND COLUMNS

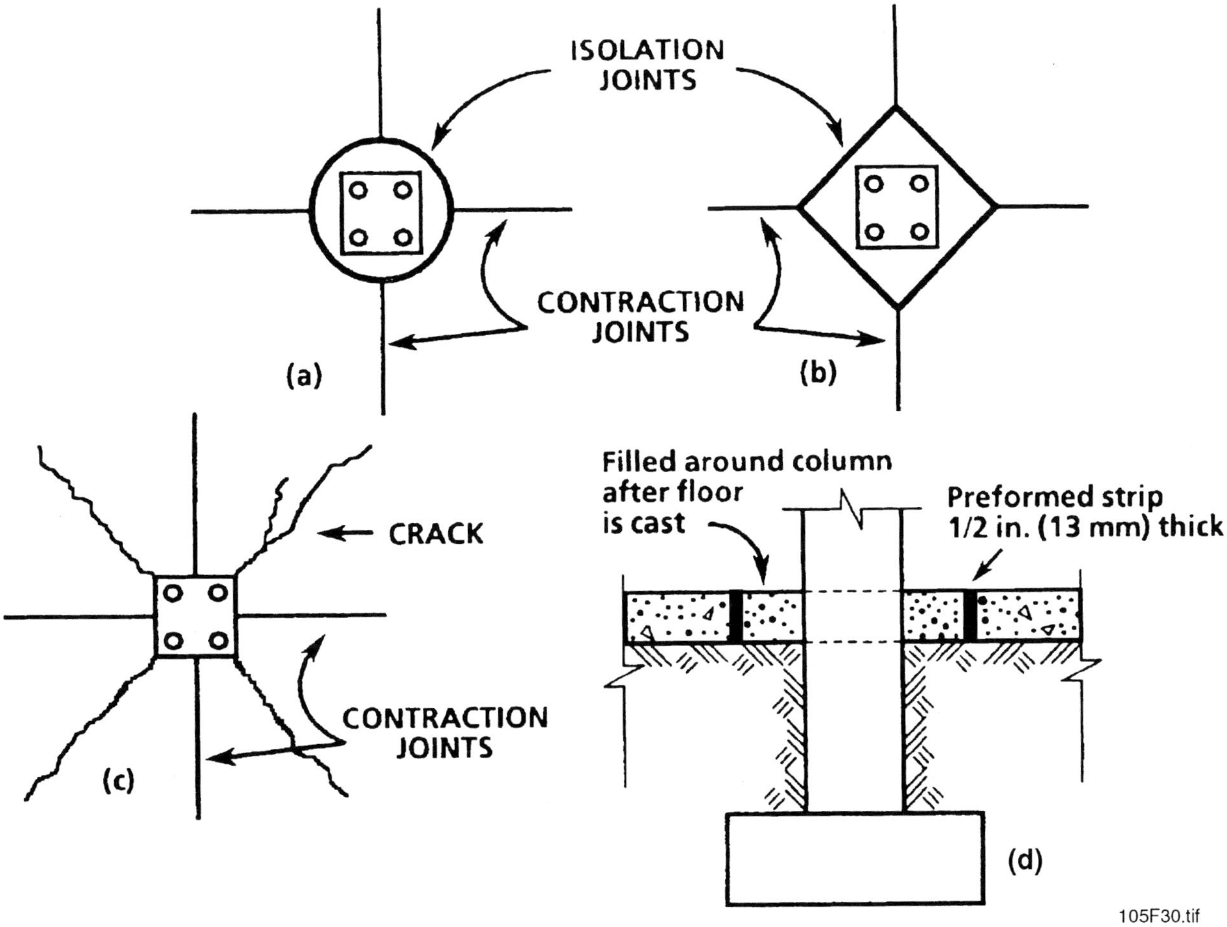

105F30.tif

TRANSPARENCY 24

FIGURE 31

TASK MODULE 23105, PREPARING FOR PLACEMENT

TYPES OF CONSTRUCTION JOINTS

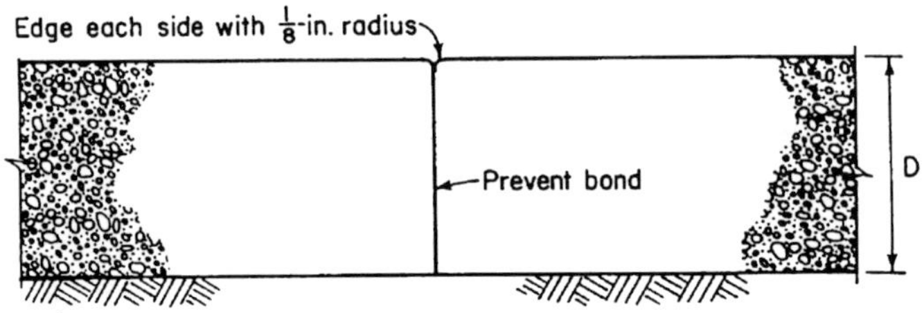

Butt-type construction joint

(a)

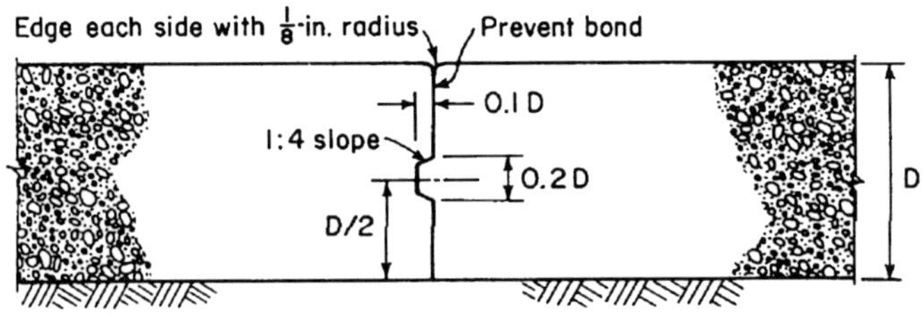

Tongue-and-groove construction joint

(b)

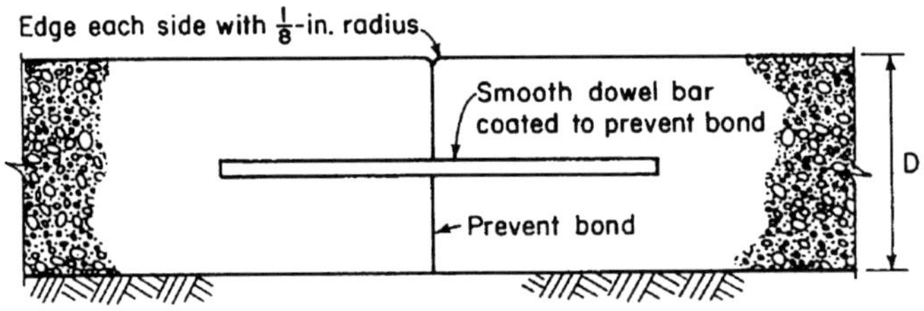

Butt-type construction joint with dowels

(c)

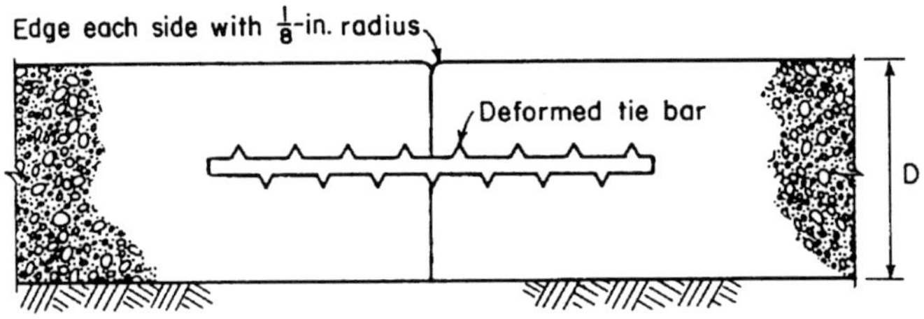

Butt-type construction joint with tie bars
(not a contraction joint)

(d)

105F31.tif

TRANSPARENCY 25

FIGURE 34

TASK MODULE 23105, PREPARING FOR PLACEMENT

STANDARD DEFORMED REINFORCING BARS

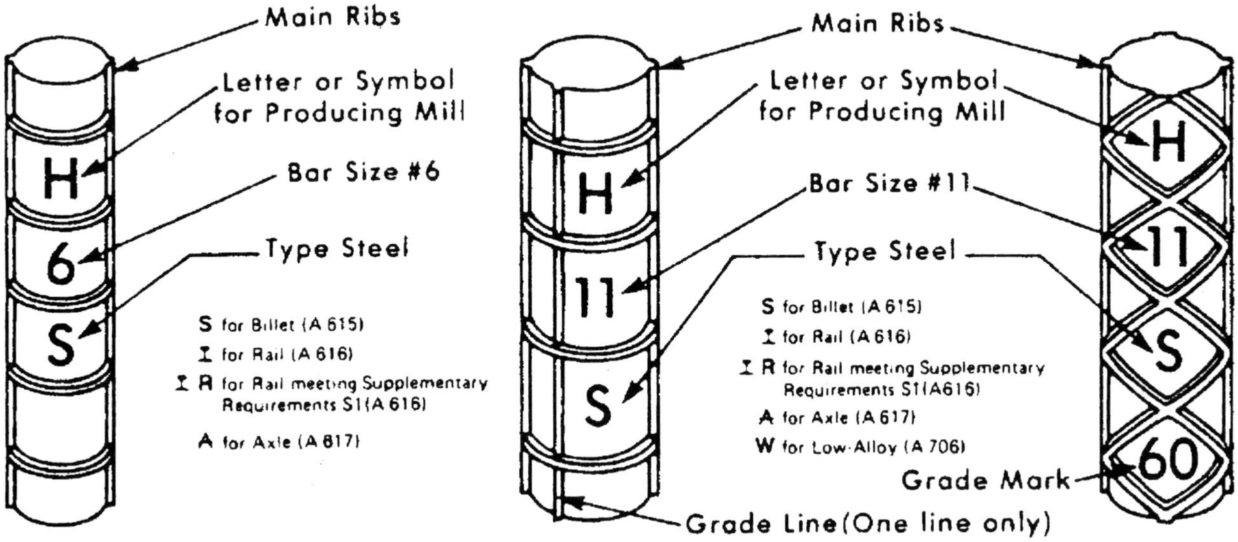

GRADE 40 AND 50

GRADE 60 AND A 706

105F34.tif

TRANSPARENCY 26

FIGURE 39

TASK MODULE 23105, PREPARING FOR PLACEMENT

WATERSTOPS

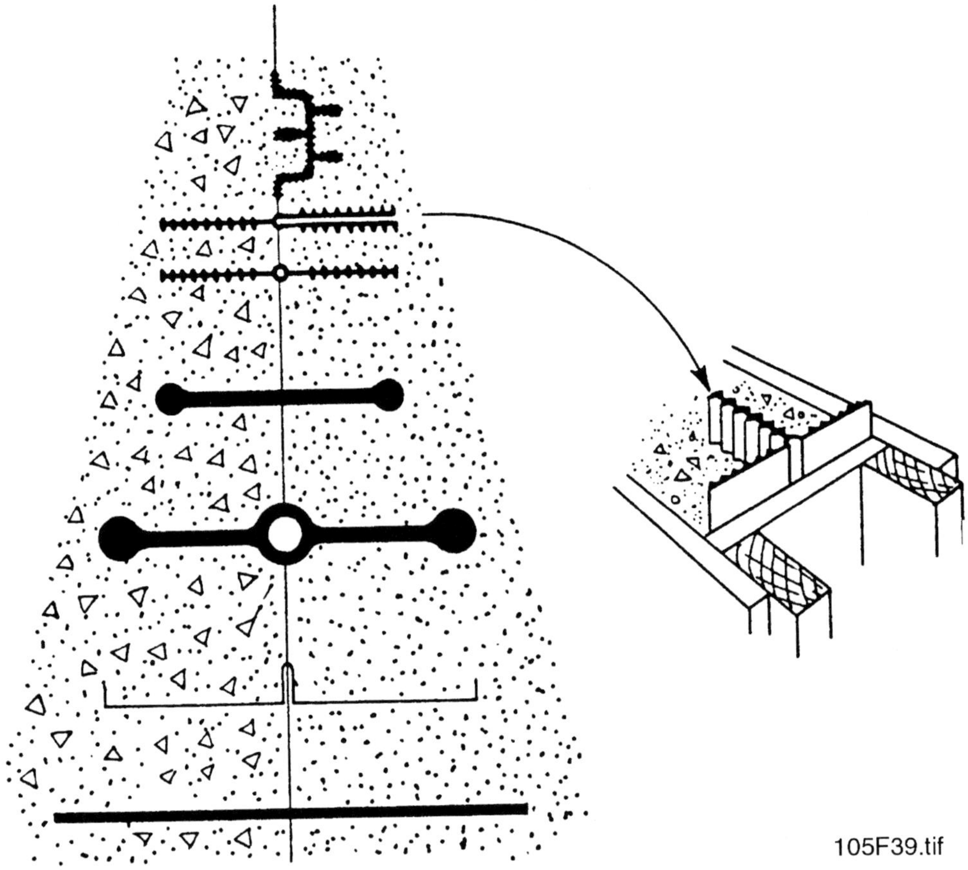

105F39.tif

TRANSPARENCY 27

FIGURE 40

TASK MODULE 23105, PREPARING FOR PLACEMENT

ACI STANDARD SPECIFICATIONS FOR TOLERANCES (EXAMPLE)

TOLERANCES

3.4 — Cross-sectional dimensions

3.4.1 *Footings*

3.4.1.1 *Horizontal dimension of formed members*
.... + 2 in.
.... − ½ in.

3.4.1.2 *Horizontal dimension of unformed members cast against soil*
2 ft. or less....................................... + 3 in.
.... − ½ in.
Greater than 2 ft. but less than 6 ft........ + 6 in.
..... − ½ in.
Over 6 ft....................................... + 12 in.
.... − ½ in.

3.4.1.3 *Vertical dimension (thickness)* − 5 percent

3.5 — Relative alignment

3.5.1 Footing side and top surfaces may slope with respect to the specified plane at a rate not to exceed the following amounts in 10 ft..............................1 in.

SECTION 4 — CAST-IN-PLACE CONCRETE FOR BUILDINGS

4.1 — Vertical alignment

4.1.1 *For heights 100 ft or less*
Lines, surfaces, and arrises.....................1 in.
Outside corner of exposed corner columns and control joint grooves in concrete exposed to view.. ½ in.

4.1.2 *For heights greater than 100 ft*
Lines, surfaces, and arrises, 1/1000 times the height but not more than.................................6 in.
Outside corner of exposed corner columns and control joint grooves in concrete, 1/2000 times the height but not more than........................3 in.

4.2 — Lateral alignment

4.2.1 *Members*...1 in.

4.2.2 *In slabs, centerline location of openings 12 in. or smaller and edge location of larger openings*.. ½ in.

4.2.3 *Sawcuts, joints, and weakened plane embedments in slabs*.. ¾ in.

4.3 — Level alignment

4.3.1 *Top of slabs:*

4.3.1.1 Elevation of slabs-on-grade........... ¾ in.

4.3.1.2 Elevation of top surfaces of formed slabs before removal of supporting shores .... ¾ in.

4.3.2 *Elevation of formed surfaces before removal of shores*.. ¾ in.

4.3.3 *Lintels, sills, parapets, horizontal grooves, and other lines exposed to view*......................... ½ in.

4.4 — Cross-sectional dimensions

4.4.1 *Members, such as columns, beams, piers, walls (thickness only), and slabs (thickness only)*
12 in. dimension or less.................... + ⅜ in.
.... − ¼ in.
More than 12 in. dimension but not over 3 ft dimension .... + ½ in.
.... − ⅜ in.
Over 3 ft dimension .... + 1 in.
.... − ¾ in.

4.5 — Relative alignment

4.5.1 *Stairs*
Difference in height between adjacent risers .... ⅛ in.
Difference in width between adjacent trends .... ¼ in.

4.5.2 *Grooves*
Specified width 2 in. or less .... ⅛ in.
Specified width more than 2 in. but not more than 12 in. .... ¼ in.

4.5.3 *Formed surfaces may slope with respect to the specified plane at a rate not to exceed the following amounts in 10 ft*

4.5.3.1 Vertical alignment of outside corner of exposed corner columns and control joint grooves in concrete exposed to view .... ¼ in.

4.5.3.2 All other conditions..................... ⅜ in.

4.5.4. *The offset between adjacent pieces of formwork facing material shall not exceed:*
Class of surface:
Class A .... ⅛ in.
Class B... ¼ in.
Class C... ½ in.
Class D ....1 in.

4.5.5 *Floor finish tolerances shall meet the requirements of either Section 4.5.6 or 4.5.7, as set forth by the specifier.*

4.5.6 *Floor finish tolerances as measured in accordance with ASTM E 1155-87 Standard Test Method for Determining Floor Flatness and Levelness Using the F-Number System (Inch-Pound Units)*

Floor profile quality classification	Minimum F_F F_L number required			
	Test area		Minimum local F number	
	Flatness F_F	Level F_L	Flatness F_F	Level F_L
Conventional				
Bullfloated	15	13	13	10
Straightedged	20	15	15	10
Flat	30	20	15	10
Very flat	50	30	25	15

4.5.6.1 The F_L levelness tolerance shall not apply to slabs placed on unshored form surfaces and/or shored form surfaces after the removal of shores. F_L levelness tolerances shall not apply to cambered or inclined surfaces and shall be measured within 72 hr after slab concrete placement.

105F40.tif

NOTE ON PERFORMANCE PROFILE TESTING

Performance Profiles are included in this Instructor's Guide in a format that can be easily photocopied for each trainee. The Profiles measure trainee competency in the tasks taught in this module.

Please note the number of tasks to be tested while teaching this module. Each trainee should be tested on the tasks listed in the Performance Profile. Before the performance testing, the instructor should brief the trainees on:

- test objectives and criteria,
- safety precautions, and
- procedures for each task to be tested.

The instructor administering the performance testing should also do the following:

- ensure that all of the needed equipment is available and operating properly;
- set up the testing stations;
- organize and administer the test in a way that allows for optimal performance;
- complete the Performance Profile Sheet for each trainee by assigning a score for each listed task;
- monitor adherence to all safety regulations and precautions;
- provide adequate supervision to prevent injuries; and
- take immediate and effective action to remedy any emergency.

ACCREDITATION TESTING

If this Performance Profile Testing is done in the National Center for Construction Education and Research Standardized Craft Training Program, the following conditions must be met:

1. The Craft Instructor must hold valid NCCER instructor certification.
2. The training must be delivered through a Training Program Sponsor recognized by the NCCER.
3. For every module, the specific performance testing must be completed to the satisfaction of the instructor.
4. The results of the testing must be recorded on the Craft Training Report Form. This form must be provided to the local Training Program Sponsor to be forwarded to the NCCER National Craft Training Registry.

Craft: Concrete Finishing	WHEELS
Task Module Number: 23105	OF
Task Module Title: Preparing for Placement	LEARNING

TRAINEE NAME: ______________________

TRAINEE SOCIAL SECURITY NUMBER: ______________________

CLASS: ______________________

TRAINING PROGRAM SPONSOR: ______________________

____

INSTRUCTOR: ______________________

Rating Levels: 1. Passed: performed task.
2. Failed: did not perform task.

Recognition: When testing for the NCCER Standardized Craft Training Program, be sure to record Performance Profile testing results on Craft Training Report Form 200 and submit the results to the Training Program Sponsor.

Objective	TASK	RATING
1, 2	1. Set a batter board.	
1, 2	2. Locate and level a form.	
1, 2, 3	3. Build a slab form to grade with dimension lumber.	
1, 2, 3	4. Compact subgrade with a shovel and come-along.	

NCCER CRAFT TRAINING USER UPDATES

The NCCER makes every effort to keep these manuals up-to-date and free of technical errors. We appreciate your help in this process. If you have an idea for improving this manual, or if you find an error, a typographical mistake, or an inaccuracy in the NCCER's Craft Training Manuals, please write us, using this form or a photocopy. Be sure to include the exact module number, page number, a description of the problem, and the correction, if possible. Your input will be brought to the attention of the Technical Review Committee. Thank you for your assistance.

Instructors – If you found that additional materials were necessary in order to teach this module effectively, please let us know so that we may include them in the Equipment/Materials list in the Instructor's Guide.

Write: Curriculum and Revision Department
National Center for Construction Education and Research
P.O. Box 141104
Gainesville, FL 32614-1104

Fax: 352-334-0932

Craft _______________ Module Name _______________

Module Number _______________ Page Number(s) _______________

Description of Problem _______________

____

____

____

(Optional) Correction of Problem _______________

____

____

(Optional) Your Name and Address _______________

____

____

Placing Concrete

Module 23106

PLACING CONCRETE

NATIONAL
CENTER FOR
CONSTRUCTION
EDUCATION AND
RESEARCH

OBJECTIVES

Upon completion of this module, the trainee will be able to:

1. Describe how concrete is conveyed and placed.
2. Draw up a pre-placement checklist.
3. Demonstrate the use of equipment and tools for placing concrete.
4. Demonstrate the process of depositing, spreading, consolidating, and striking off concrete in a form.
5. Associate trade terms with the appropriate processes and equipment.

Note to the Instructor

Before teaching this module, you should review the details in this Instructor's Guide for Equipment and Supplies, Testing, and the suggested Teaching Sequence. Be sure to allow ample time to prepare your own training plan or lesson plan and to gather all required equipment and materials.

Required Equipment and Materials

The following are required for instruction using this module:

Equipment
Overhead projector and screen
Whiteboard/chalkboard
Appropriate Personal Protective Equipment
Sandbox(es)
Wheelbarrows
Come-alongs
Square-end shovels
Poker vibrators
Straightedges
Darbies
Highway straightedges
Bullfloats
Slab forms

Materials
Trainee Task Module
Transparencies
Pencils and paper
Markers/chalk
Example placing checklist
Module Examinations
Performance Profile Sheets
Concrete
Water
Aggregate

HOW TO USE THIS INSTRUCTOR'S GUIDE

For each 2½ hour class session in this Instructor's Guide, the basic Presentation Sequence is as follows:

Introduction/Overview

Classroom, and/or Demonstration, and/or Laboratory

Class Break

Classroom, and/or Demonstration, and/or Laboratory

Summary

Suggested time periods for classroom sessions are included throughout this Instructor's Guide. These time periods should be adapted to meet local conditions and training requirements.

Each class session is presented with two columns of information. On the left side of the page, a narrow column provides suggested trainee and instructor actions, icons to call your attention to material, safety, audiovisual, or testing requirements, and space for your notes. The right-hand column provides the outline of the suggested presentation for each class session.

In this Instructor's Guide, the terms classroom, demonstration, and laboratory are defined and used as follows.

Classroom: Sessions are designed for lectures, group discussions, coaching, and additional activities. Trainees should be encouraged to actively participate.

Demonstration: Instructors will demonstrate all procedures before trainees attempt them. Instructors should make sure that trainees can point out all safety procedures during demonstrations to be assured of the proper use of equipment by trainees.

Laboratory: Instructors will facilitate all laboratory activities, coach trainees as they practice the procedures, monitor trainee progress, and provide feedback. The instructor will make sure that safety rules are followed at all times and that protective equipment is worn.

NCCER Standardized Craft Training Programs

The National Center for Construction Education and Research (NCCER) provides a standardized national program of accredited craft training. Key features of the program include instructor certification, competency-based training, and performance testing. The program provides trainees, instructors, and companies with a standard form of recognition through a National Craft Training Registry. The program is described in full in the *Guidelines for Accreditation*, published by the NCCER. For more information on standardized craft training, contact the NCCER at P.O. Box 141104, Gainesville, FL 32614-1104; or call 352-334-0911.

MODULE OVERVIEW

This module provides an overview to the methods and procedures used in concrete placing, including the different types of tools and equipment used in placing.

Prerequisites

Please see the Course Map. Prior to training with this module, it is recommended that the trainee shall have successfully completed the following modules:

Core Curricula

Concrete Finishing Level 1, Modules 23101 through 23105

Safety Considerations

Ensure that the trainees are equipped with Appropriate Personal Protective Equipment.

Teaching Time for This Module

Approximately 12½ hours or five sessions of training time are suggested to cover *Placing Concrete*. The training class session is a suggested 2½ hour time period, which includes at least one break. **You will need to adjust the time required for hands-on activities and testing based on your class size and resources.** All time periods for this module are suggested, and you will need to adapt the suggested lesson plan to meet your local conditions.

Suggested Teaching Sequence — Five 2½-Hour Sessions

Adjust your class times based on class size and resources.

Session	Topic	Trainee Module Section(s)
1	Introduction; Prepare Checklist; Delivery at Site; Discharging Concrete	1.0.0 – 4.8.0
2	Depositing, Spreading, and Consolidating; Strike Off and Smoothing Methods	5.0.0 – 6.3.3
3	Placing and Leveling Concrete Slab/Demonstration	1.0.0 – 6.3.3
4	Placing and Leveling Concrete Slab/Laboratory	1.0.0 – 6.3.3
5	Module Examination Performance Profile Examination	

Optional References for Advanced Study

This module is intended to present thorough resources for task training. The following reference works are suggested for both instructors and motivated trainees interested in further study. These are optional materials for continued education rather than for task training.

Concrete Construction and Estimating, Craig Avery, Craftsman Book Company, Carlsbad, CA, 1991.

Concrete Manual, International Conference of Building Officials, Whittier, CA, 1992.

Concrete Parking Lots: Eight Steps to Quality Construction, Videotape, Portland Cement Association, Skokie, IL.

Concrete Technology, Third Edition, George R. White, Delmar Publishers, Inc., Albany, NY, 1991.

Design and Control of Concrete Mixtures, Thirteenth Edition, Portland Cement Association, Skokie, IL, 1988.

Finishing Concrete Flatwork, Videotape 32:00 minutes, Portland Cement Association, Skokie, IL, 1991.

Guide for Concrete Floor and Slab Construction, ACI 302.1R-96, American Concrete Institute, Farmington Hills, MI, 1997.

The Contractor's Guide to Quality Construction, American Concrete Institute, Farmington Hills, MI, 1997.

COURSE MAP

This course map shows all of the modules in the first level of the Concrete Finishing curricula. The suggested training order begins at the bottom and proceeds up. Skill levels increase as a trainee advances on the course map. The training order may be adjusted by the local Training Program Sponsor.

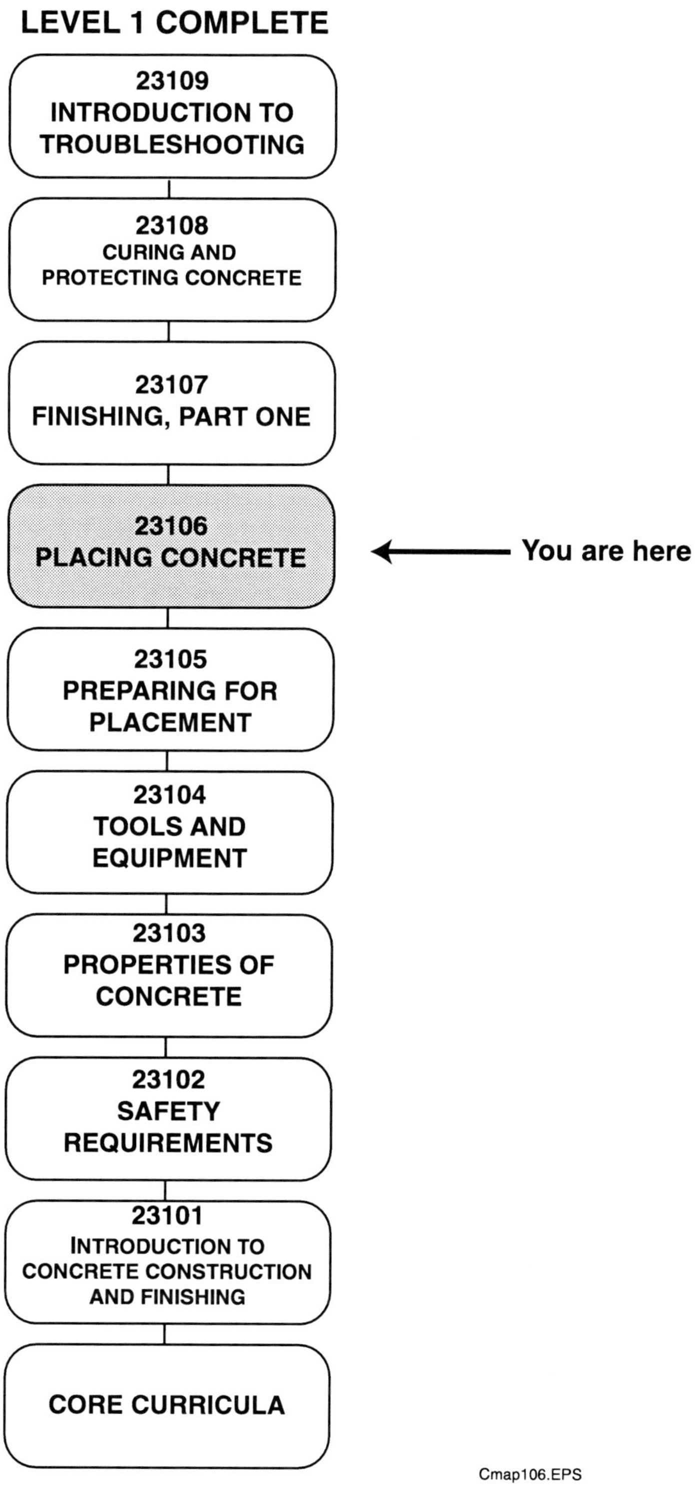

Cmap106.EPS

SESSION 1 OF 5 SESSIONS

Equipment and Materials

Ensure that all trainees have brought their Trainee Modules to class.

Trainee Task Module
Overhead projector and screen
Whiteboard/chalkboard
Markers/chalk
Transparencies
Pencils and paper
Example placing checklist

Time Schedule (150 minutes)

15 Overview
55 Classroom
15 Class Break
50 Classroom
15 Summary

Overview (15 minutes)

Show Transparency 1, Course Objectives.

Show Transparency 2, Performance Profile Tasks.

1. Take attendance and perform other administrative tasks.
2. Outline briefly the topics to be covered. Explain to the trainees that they must pass the Module Examination with a grade of at least 70 percent. Also, explain that they must pass the Performance Profile Examination by performing each task to the level of acceptance criteria.
3. Provide a brief explanation of the acceptance criteria that will be used in the Performance Profile Examination.

Classroom (55 minutes)

Show Transparency 3, Example Placement Checklist (Figure 2, Trainee Module).

Pass around an example of a placing checklist for your company.

1. Review the trade terms introduced in this module and explain them.
2. Discuss the elements that go into preparing a checklist in terms of job site specifics. Refer to the checklist in Appendix A of the Trainee Module.
   - Planning process
   - Transporting equipment
   - Site preparation and pathways
   - Crew size
   - Other variables

3. Describe the last-minute checks that need to be made.
   - Subgrade check
   - Formwork check
   - Reinforcement check
   - Materials check

   Discuss the specific procedure used by your company.

Show Transparency 4, Placing Patterns For A Large Slab (Figure 7, Trainee Module).

4. Describe placing patterns for large slabs.

Show Transparency 5, Placing Pattern For A Small Slab (Figure 8, Trainee Module).

5. Describe placing patterns for small slabs.
6. Discuss site access and travel patterns for placing equipment.

Class Break (15 minutes)

Classroom (50 minutes)

1. Ask trainees to draw up a pre-placement checklist. Make sure they include the following items:
   - Subgrade
   - Formwork
   - Reinforcement
   - Materials
2. Describe the requirements for mix delivery to the site.
   - Truck types
   - Positioning trucks
   - Truck operator responsibility

Show Transparency 6, Load Or Batch Delivery Ticket (Figure 9, Trainee Module).

3. Discuss the concrete worker's responsibility for receipt of concrete.
   - Inspecting load or batch delivery ticket
   - Visual inspection of concrete

Show Transparency 7, Common Equipment For Moving Concrete (Figure 4, Trainee Module).

4. Discuss discharging concrete by various types of equipment.
   - Chute
   - Wheelbarrow
   - Hand and motorized buggies
   - Conveyors
   - Buckets
   - Pumps
   - Dump trucks
   - Collection hoppers

Summary (15 minutes)

1. Review the materials covered in this session.
2. Answer trainee questions
3. Give assignments for the next session.

SESSION 2 OF 5 SESSIONS

Ensure that all trainees have brought their Trainee Modules to class.

Equipment and Materials

Trainee Task Module
Overhead projector and screen
Whiteboard/chalkboard
Markers/chalk
Transparencies
Pencils and paper

Time Schedule (150 minutes)

5 Overview
60 Classroom
15 Class Break
55 Classroom
15 Summary

Show Transparency 1, Course Objectives.

Show Transparency 2, Performance Profile Tasks.

Overview (5 minutes)

1. Take attendance and perform other administrative tasks.
2. Outline briefly the topics to be covered. Explain to the trainees that they must pass the Module Examination with a grade of at least 70 percent. Also, explain that they must pass the Performance Profile Examination by performing each task to the level of acceptance criteria.
3. Provide a brief explanation of the acceptance criteria that will be used in the Performance Profile Examination.

Classroom (60 minutes)

1. Review the overall construction process of laying out, forming, placing, and finishing concrete.

Show Transparency 8, Two-Lift Slab (Figure 18, Trainee Module).

2. Discuss methods for placing concrete. Explain the procedure for placing lifts. Point out the problem of bridging.
3. Discuss spreading concrete. Review the itemized list for spreading concrete.

Show Transparency 9, Spading (Figure 20, Trainee Module).

4. Discuss consolidating concrete manually by rodding, spading, and tamping.
5. Discuss the use of external mechanical vibrators.
   - Form vibrators
   - Surface vibrators
   - Vibrating screeds

Show Transparency 10, Vibrator Placement (Figure 21, Trainee Module).

6. Discuss internal vibrators. Go over the list of steps for using a spud or poker vibrator.

7. Describe how to determine if concrete is completely compacted.
8. Discuss revibration and when it should be used.

Class Break (15 minutes)

Classroom (55 minutes)

1. Describe the use of mechanical screeds with vibrating capabilities.

Show Transparency 11, Straightedge Operations.

2. Discuss straightedges. Go over straightedge operations.
3. Explain the purposes of floating the concrete.

Show Transparency 12, Highway Straightedge Operations.

4. Describe the use of a highway straightedge. Go over the procedures for using a highway straightedge.

Show Transparency 13, Bullfloat Operations.

5. Describe the use of a bullfloat. Go over the procedures for using a bullfloat.

Show Transparency 14, Darby Operations.

6. Describe darbying. Go over the procedures for using a darby.

Summary (15 minutes)

1. Review the materials covered in this session.
2. Answer trainee questions.
3. Give assignments for the next session. Explain that the next session will be a demonstration of how to place a small batch of concrete.

SESSION 3 OF 5 SESSIONS

Ensure that all trainees have brought writing materials to the site.

Equipment and Materials

Trainee Task Module
Pencils and paper
Slab forms
Concrete
Water
Wheelbarrows
Square-end shovels
Come-alongs
Poker vibrators
Straightedges
Bullfloats
Highway straightedges
Darbies
Appropriate Personal Protective Equipment

Time Schedule (150 minutes)

15 Overview
110 Demonstration
10 Class Break
15 Summary

Overview (15 minutes)

1. Take attendance and perform other administrative tasks.
2. Tell trainees to take notes and observe closely, as they will have a laboratory on these techniques for the next session.

Demonstration (110 minutes)

Make sure that all materials and equipment are available and in proper order before starting the demonstration. Concrete should be provided from a portable mixer or ready-mix truck. If ready-mix is used, you must arrange for delivery.

1. Mix concrete in portable mixer or arrange for ready-mix delivery.
2. Perform the following activities to demonstrate the proper procedures for placing concrete:
   - Place the concrete using a wheelbarrow, square-end shovel, and come-along.
   - Compact the edges of the concrete using spading.
   - Compact the remainder of the concrete using a poker vibrator.
   - Strike off the concrete with a manual screed.
   - Finish a portion of the concrete with a highway straightedge.

- Finish a portion of the concrete with a bullfloat.
- Finish a portion of the concrete with a darby.

3. Answer trainee questions about the procedures demonstrated.

Class Break (10 minutes)

Summary (15 minutes)

1. Review the methods and tools used during the placing demonstration.
2. Ask trainees to prepare for the laboratory in the next session by dressing properly and wearing the required Appropriate Personal Protective Equipment.

Ensure that all trainees have brought writing materials to the site.

SESSION 4 OF 5 SESSIONS

Equipment and Materials

Trainee Task Module
Pencils and paper
Slab forms
Concrete
Water
Wheelbarrows
Square-end shovels
Come-alongs
Straightedges
Poker vibrators
Bullfloats
Highway straightedges
Darbies
Appropriate Personal Protective Equipment

Time Schedule (150 minutes)

15 Overview
100 Laboratory
10 Class Break
25 Summary

Overview (15 minutes)

1. Take attendance and perform other administrative tasks.
2. Review the steps in the demonstration of the previous class:
    - Placing
    - Spreading
    - Compacting
    - Screeding
    - Floating
3. Outline briefly the activities to be performed during the laboratory.

Laboratory (100 minutes)

1. Ensure that all Appropriate Personal Protective Equipment is used correctly.
2. Divide trainees up into groups of two to four and have them work as a team during the placing activities.

3. Observe and monitor performance as trainees:
    - Mix concrete
    - Place concrete
    - Consolidate concrete
    - Screed concrete
    - Float concrete
4. Have trainees who have finished the placing activities begin work on the Review/Practice questions in the back of the Trainee Module.

Class Break (10 minutes)

Summary (25 minutes)

1. Review the methods and tools used during the placing activities.
2. Answer any questions about placing concrete. Discuss the Review/Practice questions.
3. Explain that the next session will be the last session and will include the Module Examination and the Performance Profile Examination.

SESSION 5 OF 5 SESSIONS

Ensure that all trainees have brought their Trainee Modules to class.

Equipment and Materials

Trainee Task Module
Pencils and paper
Slab forms
Concrete
Water
Wheelbarrows
Square-end shovels
Come-alongs
Poker vibrators
Straightedges
Bullfloats
Highway straightedges
Darbies
Appropriate Personal Protective Equipment
Module Examinations
Performance Profile Sheets

Time Schedule (150 minutes)

10 Overview
30 Module Examination
10 Class Break
100 Performance Profile Examination

Show Transparency 1, Course Objectives.

Show Transparency 2, Performance Profile Tasks.

Overview (10 minutes)

1. Take attendance and perform other administrative tasks.
2. Outline briefly the topics to be covered. Explain to the trainees that they must pass the Module Examination with a grade of at least 70 percent. Also, explain that they must pass the Performance Profile Examination by performing each task to the level of acceptance criteria.
3. Provide a brief explanation of the acceptance criteria that will be used in the Performance Profile Examination.

Module Examination (30 minutes)

1. Hand out Module Examinations to all trainees. Tell them they have 30 minutes to complete the exam.
2. Collect and grade the examinations.

TESTING

Class Break (10 minutes)

Performance Profile Examination (100 minutes)

1. Have trainees perform the following tasks to the satisfaction of the instructor:
    - Place concrete with come-alongs, wheelbarrows, and square-end shovels.
    - Consolidate placed concrete by spading and using a poker vibrator.
    - Strike off placed concrete using a manual screed.
    - Float concrete after striking off.
2. Fill out Performance Profile Sheets for each trainee.

MODULE EXAMINATIONS AND ANSWER KEYS

The masters for Module Examinations and Answer Keys are found in the booklet entitled "Concrete Finishing Level One Module Examinations" included in this Instructor's Guide package. You may also reproduce the tests using the Windows-compatible software program which may be purchased separately from Prentice Hall. If you do not have the reproducible masters in your Instructor's Guide package, or would like to order the test software, please contact Prentice Hall Publishers at 1-800-922-0579.

Be sure to keep answer keys and tests in a safe area.

OVERHEAD TRANSPARENCIES

Masters for overhead transparencies are found on the following pages. Transparencies can be made by using clear acetate sheets in a photocopy machine. This can be done at any commercial copy shop, or by following this method on your office machine:

1. Use standard 8½ × 11 clear acetate sheets, which can be obtained at office supply stores, stationery stores, or commercial copying centers.
2. Check the photocopier's manual to make sure it will take the thicker acetate sheets.
3. Remove the paper supply in the photocopying machine and replace it with a supply of acetate sheets.
4. Make photocopies from the masters on the following pages in the usual way. The material will be reproduced on the clear acetate sheets for use on any overhead projector.

TRANSPARENCY 1

TASK MODULE 23106, PLACING CONCRETE

COURSE OBJECTIVES

Upon completion of this module, the trainee will be able to:

1. **Describe how concrete is conveyed and placed.**
2. **Draw up a pre-placement checklist.**
3. **Demonstrate the use of equipment and tools for placing concrete.**
4. **Demonstrate the process of depositing, spreading, consolidating, and striking off concrete in a form.**
5. **Associate trade terms with the appropriate processes and equipment.**

TRANSPARENCY 2

TASK MODULE 23106, PLACING CONCRETE

PERFORMANCE PROFILE TASKS

1. Place concrete using a wheelbarrow, square-end shovel, and come-along.
2. Consolidate placed concrete using spading and a poker vibrator.
3. Strike off placed concrete using a manual screed.
4. Float concrete after striking off.

TRANSPARENCY 3

FIGURE 2

TASK MODULE 23106, PLACING CONCRETE

EXAMPLE PLACEMENT CHECKLIST

Concrete Placement Checklist

Job Name: ______________________ **Job No.:** ____________

Contractor: ______________________ **Contract No.:** ____________

Pour location: ______________________

Pour date: ____________ Estm CY: ____________ Estm SF: ____________

Type of structure: ______________________

DATE	QC REP	SUPT/ENG	COMPANY	ITEM OF WORK
				Layout – lane dimensions
				Subbase/fine grade – elevation – stability
				Mechanical/plumbing/fire protection – storm drain
				Electrical (ground cable, plates) – elevation – location
				Formwork (screed rails) – location – alignment – rigidity/bracing – elevation
				Embedments (guard posts, anchor bolts) – elevation – location
				Reinforcement (resteel, dowels) – spacing, elevation, clearance – lap location and length – rigidity of chairs, resteel
				Waterproofing
				Preparation for pour – clean – pumps – equipment ready – machines – screeds – vibrators
				Other

Pour start	Concrete mix	Temp Lo	Temp Hi
Pour finish	Concrete supplier	Wind	
Finish	Testing agency	Precipitation	
Curing		Sky	

Signature Concrete Contractor Carpenter Superintendent:	Date:
Signature Concrete Contractor Superintendent:	Date:
Signature Concrete Contractor QC/QA Representative:	Date:

106F02.TIF

TRANSPARENCY 4

FIGURE 7

TASK MODULE 23106, PLACING CONCRETE

PLACING PATTERNS FOR A LARGE SLAB

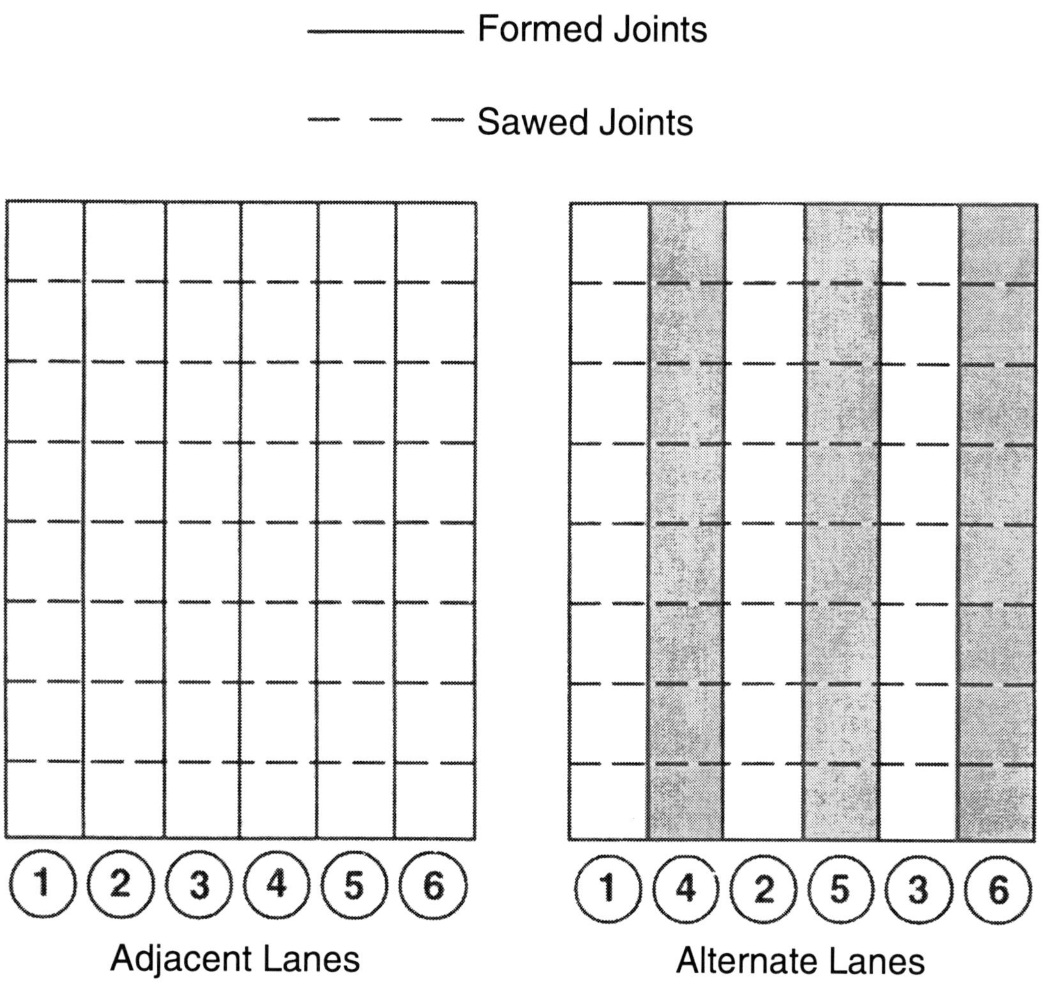

Sequence of Casting Lanes

106F07.TIF

TRANSPARENCY 5

FIGURE 8

TASK MODULE 23106, PLACING CONCRETE

PLACING PATTERN FOR A SMALL SLAB

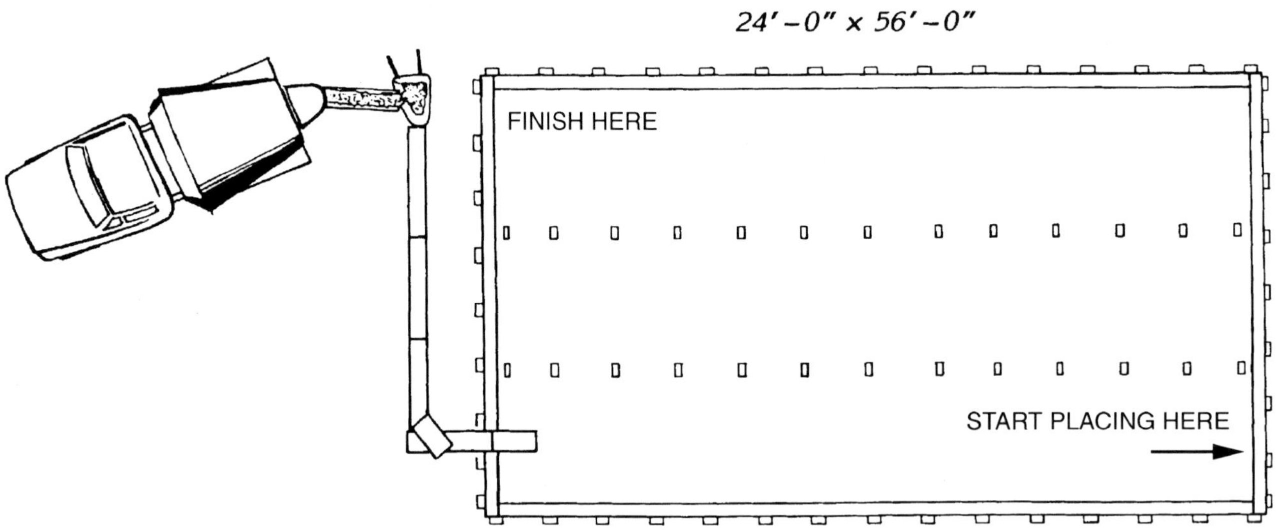

106F08.TIF

TRANSPARENCY 6

FIGURE 9

TASK MODULE 23106, PLACING CONCRETE

LOAD OR BATCH DELIVERY TICKET

Rev. 8-25-75

NAME OF CONCRETE COMPANY

Del. Ticket
Serial No. 10210

Plant No. 86-C10

Date 8-6 1975

Delivered To San Marco Construction

Address Road 860 at 50th Street

Truck No.	Time Loaded	Arrived	Discharged	Cubic Yards This Load
26	1:55 PM	*	*	8
DOT Class	DOT Mix No.	X	X	Cubic Yards Total Today
II	2001			8

Allowable Jobsite Water Addition 1½ (gals)/lbs.

Mixing Revolutions: At Plant 75 At Jobsite * (or Central Mixing Time:)

FILL OUT ON FIRST DELIVERY AND ON EACH CHANGE OF AGG. WEIGHTS	DOT Job No. 86925 -3503
Air Darex (Brand) 5.5 (Amount) oz	Retarder WRDA 79 (Brand) 45 (Amount) oz
Cement I (Type) Fla.Ptld. (Brand) 564# (Amount)	Batch Water 22 (Amount) (gals)/lbs.
Coarse Agg. 1773# (Amount) 2.0 (%Moisture)	Fine Agg. 1217# (Amount) 4.0 (%Moisture)
Flyash — (Source) (Amount)	

Issuance of this ticket constitutes certification to the accuracy of the above recorded information.

Tom Burgess
Signature of Plant Operator or Company Rep.

Water Added by Receiver of Concrete: * gals./lbs.

Received By: ______

106F09.TIF

TRANSPARENCY 7

FIGURE 4

TASK MODULE 23106, PLACING CONCRETE

COMMON EQUIPMENT FOR MOVING CONCRETE

EQUIPMENT	ADVANTAGES	DISADVANTAGES
Wheelbarrow	Versatile for small loads and short hauls where conditions are always changing	Labor intensive; may need runways to avoid segregation
Buggy	Good for short hauls where placing conditions are always changing	Labor intensive; may need runways to avoid segregation
Chute	Gravity-powered; good for shorter distances and below-grade deposit	Must be supported, may need drop chute
Conveyor	Carry horizontally up to 3,000 feet, vertically about 100 feet; adjustable	Needs drop chute to prevent segregation
Bucket and Lifter	Moved by cranes or hoists over any distance	Needs cranes or hoists for lifting
Pump	Can deliver in continuous flow with flexible pipes over obstructions and around corners	Needs continuous input of fresh, plastic mix
Cableway	Commonly used for dams and other large-scale projects taking years to build	Very expensive
Hopper	Gives on-site temporary storage; transfers to other equipment	Complicates deposit
Dump truck	Commonly used for road building; can have end, side, or bottom dump gate	Needs mixer nearby; can carry only 30 minutes
Ready-mix truck	Combines mix and transport; most commonly used; has chute attached for reach of approximately 12 feet	Time deliveries to pace of on-site work

106F04.eps

TRANSPARENCY 8

FIGURE 18

TASK MODULE 23106, PLACING CONCRETE

TWO-LIFT SLAB

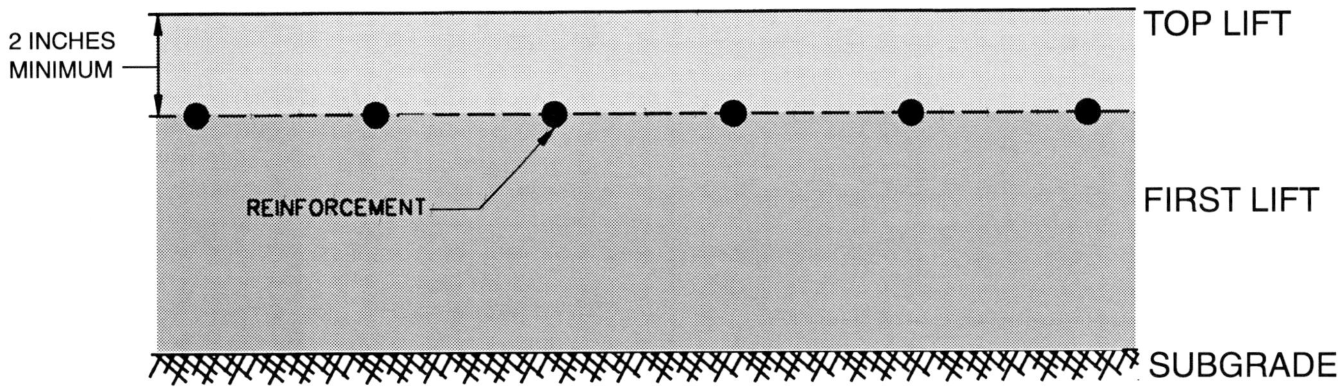

106F18.TIF

TRANSPARENCY 9

FIGURE 20

TASK MODULE 23106, PLACING CONCRETE

SPADING

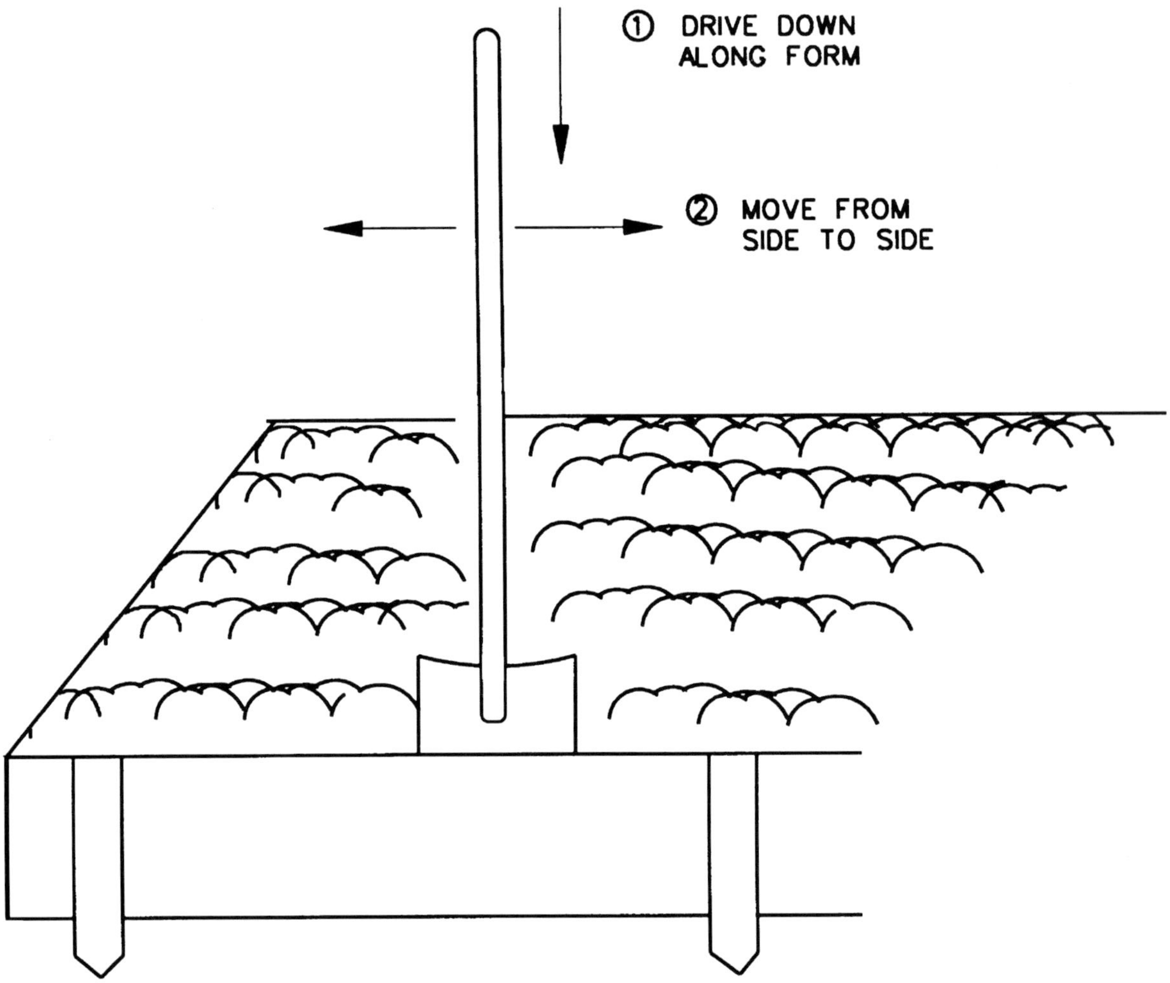

106F20.TIF

TRANSPARENCY 10

FIGURE 21

TASK MODULE 23106, PLACING CONCRETE

VIBRATOR PLACEMENT

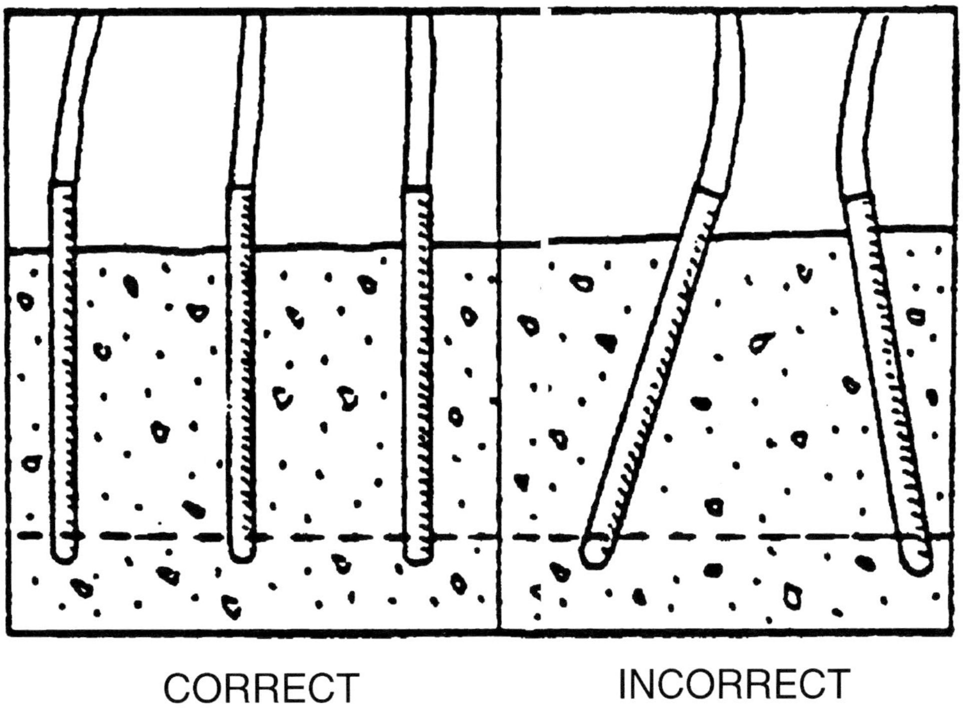

106F21.tif

TRANSPARENCY 11

TASK MODULE 23106, PLACING CONCRETE

STRAIGHTEDGE OPERATIONS

- **Pull straightedge forward in side-to-side motion.**
- **Tilt top edge 15 to 20 degrees forward.**
- **Keep bottom edge in contact with top of form.**
- **Keep a low (½ to 1 inch) surcharge.**

TRANSPARENCY 12

TASK MODULE 23106, PLACING CONCRETE

HIGHWAY STRAIGHTEDGE OPERATIONS

- **Move at 90 degrees to the direction of screeding.**
- **Pull it toward you in a continuous pull.**
- **Keep the handle low.**
- **Overlap the previous pass by ⅓ to ½ its width.**

TRANSPARENCY 13

TASK MODULE 23106, PLACING CONCRETE

BULLFLOAT OPERATIONS

- Start at 90 degrees to the direction of screeding.
- Push bullfloat away from you for first pass.
- Pull back toward you.
- Keep the handle low.
- If wood, raise the front of the blade slightly on the first pass; keep it flat on others.
- If magnesium, keep the blade flat against the surface at all times.
- Use a forward-and-backward motion to fill in holes.

TRANSPARENCY 14

TASK MODULE 23106, PLACING CONCRETE

DARBY OPERATIONS

- Move the blade in a rapid sawing motion while moving in a sweeping arc.
- Cover the surface in a regular pattern.
- Move at an angle to the path of the screed.
- Use kneeling boards to move onto the slab.

NOTE ON PERFORMANCE PROFILE TESTING

Performance Profiles are included in this Instructor's Guide in a format that can be easily photocopied for each trainee. The Profiles measure trainee competency in the tasks taught in this module.

Please note the number of tasks to be tested while teaching this module. Each trainee should be tested on the tasks listed in the Performance Profile. Before the performance testing, the instructor should brief the trainees on:

- test objectives and criteria,
- safety precautions, and
- procedures for each task to be tested.

The instructor administering the performance testing should also do the following:

- ensure that all of the needed equipment is available and operating properly;
- set up the testing stations;
- organize and administer the test in a way that allows for optimal performance;
- complete the Performance Profile Sheet for each trainee by assigning a score for each listed task;
- monitor adherence to all safety regulations and precautions;
- provide adequate supervision to prevent injuries; and
- take immediate and effective action to remedy any emergency.

ACCREDITATION TESTING

If this Performance Profile Testing is done in the National Center for Construction Education and Research Standardized Craft Training Program, the following conditions must be met:

1. The Craft Instructor must hold valid NCCER instructor certification.
2. The training must be delivered through a Training Program Sponsor recognized by the NCCER.
3. For every module, the specific performance testing must be completed to the satisfaction of the instructor.
4. The results of the testing must be recorded on the Craft Training Report Form. This form must be provided to the local Training Program Sponsor to be forwarded to the NCCER National Craft Training Registry.

Craft: Concrete Finishing

Task Module Number: 23106

Task Module Title: Placing Concrete

WHEELS OF LEARNING

TRAINEE NAME: ______________________________

TRAINEE SOCIAL SECURITY NUMBER: ______________________________

CLASS: ______________________________

TRAINING PROGRAM SPONSOR: ______________________________

____

INSTRUCTOR: ______________________________

Rating Levels: 1. Passed: performed task.
2. Failed: did not perform task.

Recognition: When testing for the NCCER Standardized Craft Training Program, be sure to record Performance Profile testing results on Craft Training Report Form 200 and submit the results to the Training Program Sponsor.

Objective	TASK	RATING
3, 4	1. Place concrete using a wheelbarrow, square-end shovel, and come-along.	
3, 4	2. Consolidate placed concrete using spading and a poker vibrator.	
3, 4	3. Strike off placed concrete using a manual screed.	
3, 4	4. Float concrete after striking off.	

NCCER CRAFT TRAINING USER UPDATES

The NCCER makes every effort to keep these manuals up-to-date and free of technical errors. We appreciate your help in this process. If you have an idea for improving this manual, or if you find an error, a typographical mistake, or an inaccuracy in the NCCER's Craft Training Manuals, please write us, using this form or a photocopy. Be sure to include the exact module number, page number, a description of the problem, and the correction, if possible. Your input will be brought to the attention of the Technical Review Committee. Thank you for your assistance.

Instructors – If you found that additional materials were necessary in order to teach this module effectively, please let us know so that we may include them in the Equipment/Materials list in the Instructor's Guide.

Write: Curriculum and Revision Department
National Center for Construction Education and Research
P.O. Box 141104
Gainesville, FL 32614-1104

Fax: 352-334-0932

Craft ____________ Module Name ____________

Module Number ____________ Page Number(s) ____________

Description of Problem ____________

____

____

____

(Optional) Correction of Problem ____________

____

____

(Optional) Your Name and Address ____________

____

____

Finishing, Part I

Module 23107

Concrete Finishing Instructor's Guide Task Module 23107

NATIONAL CENTER FOR CONSTRUCTION EDUCATION AND RESEARCH

FINISHING, PART ONE

OBJECTIVES

Upon completion of this module, the trainee will be able to:

1. Describe the basic finishing process.
2. Use the following finishing hand tools: float, edger, groover, and trowel.
3. Mark and cut joints with a saw.
4. Apply a broom finish.
5. Apply a rubbing finish.
6. Associate trade terms with the appropriate processes and equipment.

Note to the Instructor

Before teaching this module, you should review the details in this Instructor's Guide for Equipment and Supplies, Testing, and the suggested Teaching Sequence. Be sure to allow ample time to prepare your own training plan or lesson plan and to gather all required equipment and materials.

Required Equipment and Materials

The following are required for instruction using this module:

Equipment

Overhead projector and screen
Whiteboard/chalkboard
Appropriate Personal Protective Equipment
Sandboxes (4 ft. × 4 ft.)
Hammers
Come-alongs
Square-end shovels
Straightedges
Darbies
Bullfloats
Hand floats
Edgers
Groovers
Trowels
Concrete saw(s)
Rubbing stone(s)
Chalk lines
Steel tape
Brooms
Wheelbarrows
Power saws

Materials

Trainee Task Module
Transparencies
Markers/chalk
Pencils and paper
Nails
Concrete (ready-mixed recommended)
Sand
Lime
Water
Dimensional lumber slab form
Burlap
Grout mixture
Marking pencils
Module Examinations
Performance Profile Sheets

HOW TO USE THIS INSTRUCTOR'S GUIDE

For each 2½ hour class session in this Instructor's Guide, the basic Presentation Sequence is as follows:

Introduction/Overview

Classroom, and/or Demonstration, and/or Laboratory

Class Break

Classroom, and/or Demonstration, and/or Laboratory

Summary

Suggested time periods for classroom sessions are included throughout this Instructor's Guide. These time periods should be adapted to meet local conditions and training requirements.

Each class session is presented with two columns of information. On the left side of the page, a narrow column provides suggested trainee and instructor actions, icons to call your attention to material, safety, audiovisual, or testing requirements, and space for your notes. The right-hand column provides the outline of the suggested presentation for each class session.

In this Instructor's Guide, the terms classroom, demonstration, and laboratory are defined and used as follows.

Classroom: Sessions are designed for lectures, group discussions, coaching, and additional activities. Trainees should be encouraged to actively participate.

Demonstration: Instructors will demonstrate all procedures before trainees attempt them. Instructors should make sure that trainees can point out all safety procedures during demonstrations to be assured of the proper use of equipment by trainees.

Laboratory: Instructors will facilitate all laboratory activities, coach trainees as they practice the procedures, monitor trainee progress, and provide feedback. The instructor will make sure that safety rules are followed at all times and that protective equipment is worn.

NCCER Standardized Craft Training Programs

The National Center for Construction Education and Research (NCCER) provides a standardized national program of accredited craft training. Key features of the program include instructor certification, competency-based training, and performance testing. The program provides trainees, instructors, and companies with a standard form of recognition through a National Craft Training Registry. The program is described in full in the *Guidelines for Accreditation*, published by the NCCER. For more information on standardized craft training, contact the NCCER at P.O. Box 141104, Gainesville, FL 32614-1104; or call 352-334-0911.

MODULE OVERVIEW

In this module, basic finishing techniques for slabs and other horizontal structures are described. Use of floats, trowels, edgers, and groovers is explained and demonstrated. Requirements for cutting joints using different types of saws are presented. Hands-on practice is provided for finishing concrete slabs.

Prerequisites

Please see the Course Map. Prior to training with this module, it is recommended that the trainee shall have successfully completed the following modules:

Core Curricula

Concrete Finishing Level 1, Modules 23101 through 23106

Safety Considerations

Ensure that the trainees are equipped with Appropriate Personal Protective Equipment.

Teaching Time for This Module

Approximately 20 hours or eight sessions of training time are suggested to cover *Finishing, Part One*. The training class session is a suggested 2½ hour time period, which includes at least one break. **You will need to adjust the time required for hands-on activities and testing based on your class size and resources.** All time periods for this module are suggested, and you will need to adapt the suggested lesson plan to meet your local conditions.

Suggested Teaching Sequence — Eight 2½-Hour Sessions

Adjust your class times based on class size and resources.

Session	Topic	Trainee Module Section(s)
1	Introduction; Basic Finishing Processes; Edging Concrete; Jointing Concrete; Finishing Techniques; Surface Treatments	1.0.0 – 6.6.2
2	Floating Techniques – Demonstration/Laboratory	
3	Edging, Grooving, and Troweling – Demonstration/Laboratory	
4	Finishing Small Slab – Demonstration/Laboratory	
5	Finishing Small Slab – Laboratory	
6	Cutting Joints – Demonstration/Laboratory	
7	Finishing – Demonstration/Laboratory	
8	Module Examination Performance Profile Examination	

Optional References for Advanced Study

This module is intended to present thorough resources for task training. The following reference works are suggested for both instructors and motivated trainees interested in further study. These are optional materials for continued education rather than for task training.

Concrete Construction and Estimating, Craig Avery, Craftsman Book Company, Carlsbad, CA, 1991.

Concrete Manual, International Conference of Building Officials, Whittier, CA, 1992.

Concrete Parking Lots: Eight Steps to Quality Construction, Videotape, Portland Cement Association, Skokie, IL.

Concrete Technology, Third Edition, George R. White, Delmar Publishers, Inc., Albany, NY, 1991.

Design and Control of Concrete Mixtures, Thirteenth Edition, Portland Cement Association, Skokie, IL, 1988.

Finishing Concrete Flatwork, Videotape 32:00 minutes, Portland Cement Association, Skokie, IL, 1991.

Guide for Concrete Floor and Slab Construction, ACI 302.1R-96, American Concrete Institute, Farmington Hills, MI, 1997.

The Contractor's Guide to Quality Construction, American Concrete Institute, Farmington Hills, MI, 1997.

COURSE MAP

This course map shows all of the modules in the first level of the Concrete Finishing curricula. The suggested training order begins at the bottom and proceeds up. Skill levels increase as a trainee advances on the course map. The training order may be adjusted by the local Training Program Sponsor.

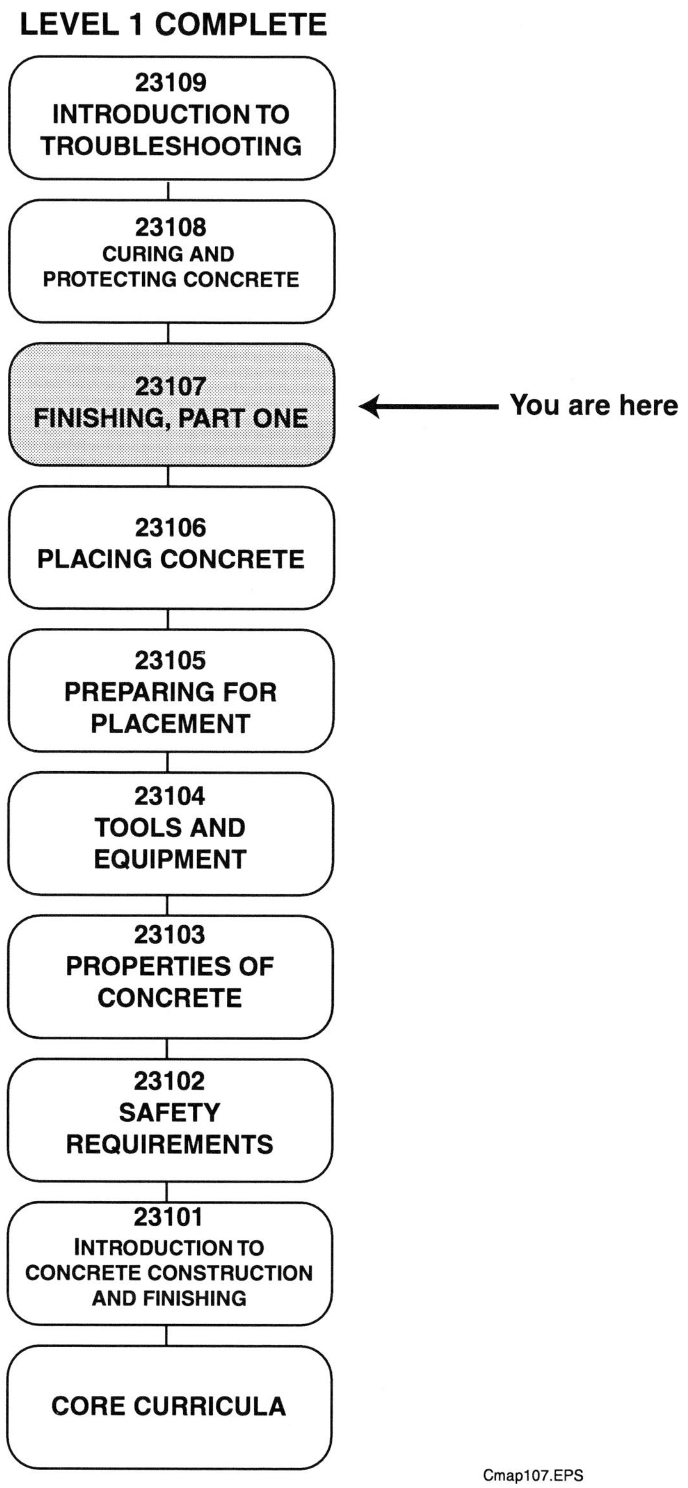

Cmap107.EPS

SESSION 1 OF 8 SESSIONS

Equipment and Materials

Trainee Task Module
Overhead projector and screen
Whiteboard/chalkboard
Markers/chalk
Transparencies
Pencils and paper

Ensure that all trainees have brought their Trainee Modules to class.

Time Schedule (150 minutes)

15 Overview
55 Classroom
15 Class Break
55 Classroom
10 Summary

Overview (15 minutes)

Show Transparency 1, Course Objectives.

Show Transparency 2, Performance Profile Tasks.

1. Take attendance and perform other administrative tasks.
2. Outline briefly the topics to be covered. Explain to the trainees that they must pass the Module Examination with a grade of at least 70 percent. Also, explain that they must pass the Performance Profile Examination by performing each task to the level of acceptance criteria.
3. Provide a brief explanation of the acceptance criteria that will be used in the Performance Profile Examination.

Classroom (55 minutes)

Use the whiteboard to help explain the trade terms.

1. Review the trade terms introduced in this module and briefly explain them.
2. Give trainees an overview of the basic finishing activities. Include:
   - Bullfloating
   - Bleed water removal
   - Edging
   - Jointing
   - Floating
   - Troweling
   - Surface treatment

AUDIO/VISUAL

Show Transparency 3, Edging Techniques.

3. Explain the procedures used in edging.
   - Purpose
   - Techniques
4. Review the purpose of jointing and the types of joints used.
5. Describe resetting the construction joint edge.

Show Transparency 4, Hand Jointing Techniques.

6. Describe grooving joints in fresh concrete. Explain hand techniques.
7. Explain the overall requirements for sawing joints in concrete:
   - Conventional wet and dry cutting
   - Early entry dry cutting

Show Transparency 5, Jointing With A Power Saw.

8. Describe jointing with a saw.
9. Describe the different types of joint sealers used in your area. Explain how they are installed.

Class Break (15 minutes)

Classroom (55 minutes)

1. Explain the processes for hand floating concrete.

Show Transparency 6, Using A Darby.

   - Using a darby

Show Transparency 7, Using A Hand Float.

   - Using a float
   - Using knee boards

Show Transparency 8, Power Floating.

2. Discuss power floating:
   - Methods

Show Transparency 9, Power Floating Path (Figure 16, Trainee Module).

   - Floating path

Show Transparency 10, Using A Hand Trowel.

3. Describe the process of hand troweling concrete.

Show Transparency 11, Floating and Troweling.

4. Describe how floating and troweling can be done in the same operation.

AUDIO/VISUAL

Show Transparency 12, Power Troweling.

5. Explain the operation of a power trowel. Describe the patterns that should be followed to trowel the surface.
   - Basics of operation
   - Guiding the trowel
   - Requirements for the direction of the trowel

Show Transparency 13, Trowel Path (Figure 22, Trainee Module).

6. Explain the requirements for tolerances and flatness in slabs.
   - Straightedge method
   - F-Number method

Show Transparency 14, F-Number Use Guide (Figure 23, Trainee Module).

7. Describe the options available for surface treatments.
8. Describe interior surface treatment procedures.
   - Plain
   - Heavy duty
9. Describe exterior surface treatment procedures.
   - Air-entrained surface
   - Plain treatment
10. Describe broom finishing.
11. Describe safety grooved surface treatments.

Show Transparency 15, Rubbing (Figure 26, Trainee Module).

12. Discuss rubbing.
    - Sack rubbing
    - Stone rubbing

Summary (10 minutes)

1. Review the materials covered in this session.
2. Answer trainee questions
3. Give assignments for the next session. Tell trainees that the remainder of the sessions will be a combination of demonstrations and laboratories. Remind them to bring Appropriate Personal Protective Equipment.

SESSION 2 OF 8 SESSIONS

Equipment and Materials

Ensure that all trainees have brought their Trainee Modules to class.

- Trainee Task Module
- Whiteboard/chalkboard
- Markers/chalk
- Pencils and paper
- Sandboxes (4 ft. × 4 ft.)
- Sand
- Lime
- Water
- Bullfloats
- Darbies
- Straightedges
- Hand floats
- Appropriate Personal Protective Equipment

The sandboxes should be constructed with 2 x 4 dimensional lumber on a flat surface. The material to use with the demonstration and laboratory consists of sand, lime, and water mixture which will simulate the concrete mix.

Time Schedule (150 minutes)

10 Overview
45 Demonstration
15 Class Break
70 Laboratory
10 Summary

Overview (10 minutes)

1. Take attendance and perform other administrative tasks.
2. Outline briefly the topics to be covered in the demonstration and laboratory.

Demonstration (45 minutes)

Make sure sandboxes are constructed prior to beginning the session. Each box should be approximately 4 feet by 4 feet. Have a sufficient number of boxes to accommodate the size of the class.

1. Using the sandbox and sand mixture, demonstrate the proper techniques for using the following hand tools:
   - Straightedge
   - Bullfloat
   - Darby
   - Hand float
2. Answer trainee questions.

Class Break (15 minutes)

Laboratory (70 minutes)

1. Coach and monitor trainees as they practice using the following hand tools in the sandboxes:
    - Straightedge
    - Bullfloat
    - Darby
    - Hand float
2. Ensure that all safety regulations are followed.

Summary (10 minutes)

1. Review the materials covered in this session.
2. Answer trainee questions.

MATERIALS

Ensure that all trainees have brought their Trainee Modules to class.

SESSION 3 OF 8 SESSIONS

Equipment and Materials

Trainee Task Module
Whiteboard/chalkboard
Markers/chalk
Pencils and paper
Sandboxes (4 ft. × 4 ft.)
Steel tape
Sand
Lime
Water
Edgers
Groovers
Trowels
Chalk lines
Marking pencils
Appropriate Personal Protective Equipment

Time Schedule (150 minutes)

10 Overview
45 Demonstration
15 Class Break
70 Laboratory
10 Summary

Overview (10 minutes)

1. Take attendance and perform other administrative tasks.
2. Outline briefly the topics to be covered in the demonstration and laboratory.

Demonstration (45 minutes)

1. Using the sandbox and sand mixture, demonstrate the proper technique for marking joints with a marker and chalk line.
2. Using the sandbox and sand mixture, demonstrate the proper techniques for using the following hand tools:
   - Edger
   - Groover
   - Trowel
3. Answer trainee questions about the techniques.

Class Break (15 minutes)

Laboratory (70 minutes)

SAFETY

1. Coach and monitor trainees as they use a sandbox and sand mixture to perform the following tasks:
   - Mark joints
   - Use the edger
   - Use the groover
   - Use the trowel
2. Ensure that all Appropriate Personal Protective Equipment is used correctly.

Summary (10 minutes)

1. Review the materials covered in this session.
2. Answer trainee questions about the techniques.

SESSION 4 OF 8 SESSIONS

Ensure that all trainees have brought their Trainee Modules to class.

Construct the forms prior to beginning the session. Make sure concrete has been ordered for use in the placing and finishing demonstration and laboratory practice.

Equipment and Materials

Trainee Task Module
Whiteboard/chalkboard
Markers/chalk
Dimensional lumber slab form
Concrete (ready-mixed recommended)
Wheelbarrows
Square-end shovels
Come-alongs
Straightedges
Bullfloats
Darbies
Edgers
Groovers
Chalk lines and marking pencils
Floats
Trowels
Brooms
Appropriate Personal Protective Equipment

Time Schedule (150 minutes)

5 Overview
55 Demonstration
15 Class Break
70 Laboratory
5 Summary

Overview (5 minutes)

1. Take attendance and perform other administrative tasks.
2. Review techniques that will be demonstrated and practiced during this session:
   - Place and compact concrete
   - Strike off with a straightedge
   - Bullfloat
   - Remove bleed water
   - Darby
   - Edge
   - Mark joints
   - Groove joints

- Hand float
- Hand trowel
- Broom finish

Demonstration (55 minutes)

1. Using premade forms for slabs and mountable curb sections, demonstrate the following tasks:
   - Place and compact concrete
   - Strike off with a straightedge
   - Bullfloat
   - Remove bleed water
   - Darby
   - Edge
   - Mark joints
   - Groove joints
   - Hand float
   - Hand trowel
   - Broom finish
2. Answer trainee questions.

Class Break (15 minutes)

Laboratory (70 minutes)

1. Coach and monitor trainees as they use the forms for slabs and mountable curb sections to perform the following tasks:
   - Place and compact concrete
   - Strike off with a straightedge
   - Bullfloat
   - Remove bleed water
   - Darby
   - Edge
   - Mark joints
   - Groove joints
   - Hand float
   - Hand trowel
   - Broom finish
2. Ensure that all Appropriate Personal Protective Equipment is used correctly.

Summary (5 minutes)

1. Review the tasks performed by the trainees and answer any questions about the procedures and techniques.
2. Explain that the next session will be a laboratory to give trainees time to work on finishing techniques.

SESSION 5 OF 8 SESSIONS

Ensure that all trainees have brought their Trainee Modules to class.

Make sure sufficient forms are available for practice.

Make sure a sufficient quantity of concrete has been ordered for use with each form.

Equipment and Materials

Trainee Task Module
Whiteboard/chalkboard
Markers/chalk
Dimension lumber slab forms
Concrete (ready-mixed recommended)
Wheelbarrows
Square-end shovels
Come-alongs
Straightedges
Bullfloats
Darbies
Edgers
Groovers
Chalk lines and marking pencils
Floats
Trowels
Brooms
Steel tape
Appropriate Personal Protective Equipment

Time Schedule (150 minutes)

10 Overview
125 Laboratory
15 Summary

Overview (10 minutes)

1. Take attendance and perform other administrative tasks.
2. Review the steps in the demonstration of the previous session.

Laboratory (125 minutes)

1. Observe and monitor trainees as they practice performing the following tasks:
   - Placing and compacting concrete
   - Screeding
   - Bullfloating
   - Darbying
   - Edging

- Marking and grooving joints
- Floating
- Troweling
- Brooming

2. Ensure that all Appropriate Personal Protective Equipment is used correctly.

Summary (15 minutes)

1. Review the techniques and procedures used in finishing.
2. Answer any questions about finishing techniques.

SESSION 6 OF 8 SESSIONS

Ensure that all trainees have brought their Trainee Modules to class.

Make sure concrete has hardened enough to properly use a power saw.

Equipment and Materials

Trainee Task Module
Whiteboard/chalkboard
Markers/chalk
Previously finished concrete slabs (from Session 5)
Steel tape
Chalk lines
Marking pencils
Power saws
Appropriate Personal Protective Equipment

Time Schedule (150 minutes)

5 Overview
30 Demonstration
15 Class Break
80 Laboratory
20 Summary

Overview (5 minutes)

1. Take attendance and perform other administrative tasks.
2. Review procedures for cutting joints with a power saw.

Demonstration (30 minutes)

1. Demonstrate marking joints with a marking pencil and chalk line.
2. Demonstrate cutting joints with a power saw. Use equipment most commonly used by your company.

Class Break (15 minutes)

Laboratory (80 minutes)

1. Coach and monitor trainees as they:
   - Mark joints
   - Cut joints with power saws
2. Ensure that all safety procedures are followed.

Summary (20 minutes)

1. Answer trainee questions about cutting joints with power saws.
2. Review activities performed to date.

SESSION 7 OF 8 SESSIONS

Ensure that all trainees have brought their Trainee Modules to class.

Prepare grout mixture before the session begins so it is available for the demonstration.

Equipment and Materials

Trainee Task Module
Previously finished concrete slabs (from Session 6)
Grout mixture
Rubbing stones
Floats and burlap
Appropriate Personal Protective Equipment

Time Schedule (150 minutes)

5 Overview
40 Demonstration
15 Class Break
70 Laboratory
20 Summary

Overview (5 minutes)

1. Take attendance and perform other administrative tasks.
2. Review tasks to be covered in this session.

Demonstration (40 minutes)

1. Demonstrate rubbing with burlap fabric and a float.
2. Demonstrate rubbing with a rubbing stone.

Class Break (15 minutes)

Laboratory (70 minutes)

1. Coach and monitor trainees as they:
   - Apply a sack rubbing finish
   - Apply a stone rubbing finish
2. Ensure that all safety procedures are followed.
3. Have trainees complete the review/practice questions found in the Trainee Module.

Summary (20 minutes)

1. Review activities performed to date.
2. Answer trainee questions about sack rubbing and stone rubbing.
3. Review answers to the review/practice questions.
4. Explain that the next session will be the last session for this module and that there will be a Module Examination and Performance Profile Examination.

SESSION 8 OF 8 SESSIONS

Equipment and Materials

Ensure that all trainees have brought their Trainee Modules to class.

Trainee Task Module
Overhead projector and screen
Transparencies
Pencils and paper
Dimensional lumber slab forms
Concrete (ready-mixed recommended)
Wheelbarrows
Square-end shovels
Come-alongs
Straightedges
Bullfloats
Darbies
Edgers
Marking pencils
Chalk lines
Hand floats
Trowels
Brooms
Appropriate Personal Protective Equipment
Module Examinations
Performance Profile Sheets

Time Schedule (150 minutes)

5	Overview
30	Module Examination
15	Class Break
100	Performance Profile Examination

Overview (5 minutes)

Show Transparency 1, Course Objectives.

Show Transparency 2, Performance Profile Tasks.

1. Take attendance and perform other administrative tasks.
2. Outline briefly the topics to be covered. Explain to the trainees that they must pass the Module Examination with a grade of at least 70 percent. Also, explain that they must pass the Performance Profile Examination by performing each task to the level of acceptance criteria.
3. Provide a brief explanation of the acceptance criteria that will be used in the Performance Profile Examination.

TESTING

Module Examination (30 minutes)

1. Hand out Module Examinations to all trainees. Tell them they have 30 minutes to complete the exam.
2. Collect and grade the examinations.

Class Break (15 minutes)

Performance Profile Examination (100 minutes)

1. Have trainees perform the following tasks to the satisfaction of the instructor:
   - Hand float, edge, groove, and trowel a 4 x 4 foot concrete slab.
   - Apply a broom finish to a slab.
   - Mark joints and cut them with a power saw.
   - Apply a rubbing finish to a slab.
2. Fill out Performance Profile Sheets for each trainee.

MODULE EXAMINATIONS AND ANSWER KEYS

The masters for Module Examinations and Answer Keys are found in the booklet entitled "Concrete Finishing Level One Module Examinations" included in this Instructor's Guide package. You may also reproduce the tests using the Windows-compatible software program which may be purchased separately from Prentice Hall. If you do not have the reproducible masters in your Instructor's Guide package, or would like to order the test software, please contact Prentice Hall Publishers at 1-800-922-0579.

Be sure to keep answer keys and tests in a safe area.

OVERHEAD TRANSPARENCIES

Masters for overhead transparencies are found on the following pages. Transparencies can be made by using clear acetate sheets in a photocopy machine. This can be done at any commercial copy shop, or by following this method on your office machine:

1. Use standard 8½ x 11 clear acetate sheets, which can be obtained at office supply stores, stationery stores, or commercial copying centers.
2. Check the photocopier's manual to make sure it will take the thicker acetate sheets.
3. Remove the paper supply in the photocopying machine and replace it with a supply of acetate sheets.
4. Make photocopies from the masters on the following pages in the usual way. The material will be reproduced on the clear acetate sheets for use on any overhead projector.

TRANSPARENCY 1

TASK MODULE 23107, FINISHING, PART ONE

COURSE OBJECTIVES

Upon completion of this module, the trainee will be able to:

1. **Describe the basic finishing process.**
2. **Use the following hand tools:**
   - **Float**
   - **Edger**
   - **Groover**
   - **Trowel**
3. **Mark and cut joints with a saw.**
4. **Apply a broom finish.**
5. **Apply a rubbing finish.**
6. **Associate trade terms with the appropriate processes and equipment.**

TRANSPARENCY 2

TASK MODULE 23107, FINISHING, PART ONE

PERFORMANCE PROFILE TASKS

1. Hand float, edge, groove, and trowel a 4×4 foot concrete slab.
2. Apply a broom finish to a slab.
3. Mark joints and cut them with a power saw.
4. Apply a rubbing finish to a slab.

TRANSPARENCY 3

TASK MODULE 23107, FINISHING, PART ONE

EDGING TECHNIQUES

- Cut the edge with a margin trowel.
- Insert the edger into the cut.
- Move forward, with the front of the edger tilted slightly upward.
- Lift up and reinsert the edger.
- Move backward, with the back of the edger tilted slightly upward.

TRANSPARENCY 4

TASK MODULE 23107, FINISHING, PART ONE

HAND JOINTING TECHNIQUES

- Mark the location.
- Set the cutting guide.
- Score with a pointing trowel.
- Reset the cutting guide.
- Insert the groover into the score and flush with the edge of the cutting guide.
- Push the groover forward.
- Clean the bit and move it backward.

TRANSPARENCY 5

TASK MODULE 23107, FINISHING, PART ONE

JOINTING WITH A POWER SAW

- Start when the concrete is firm—leaving less than ¼ of an inch impression.
- Mark the joints.
- Use Appropriate Personal Protective Equipment.
- Use the cutting guide.
- Move the saw slowly and carefully.
- Practice power tool safety.

TRANSPARENCY 6

TASK MODULE 23107, FINISHING, PART ONE

USING A DARBY

- Move the blade in a rapid sawing motion while also moving in an arc.
- Cover the surface in a regular pattern.
- Darby at an angle to the screed path.

TRANSPARENCY 7

TASK MODULE 23107, FINISHING, PART ONE

USING A HAND FLOAT

- Support yourself by leaning on a trowel.
- Keep the float *flat* on the surface.
- Move the float in an arc.

TRANSPARENCY 8

TASK MODULE 23107, FINISHING, PART ONE

POWER FLOATING

- Check the firmness of the slab—leaving less than ¼ of an inch impression.
- Check the equipment and blades.
- Adjust the blades flat to the surface.
- Use Appropriate Personal Protective Equipment.
- Follow the float path.
- Keep the float moving.

TRANSPARENCY 9

FIGURE 16

TASK MODULE 23107, FINISHING, PART ONE

POWER FLOATING PATH

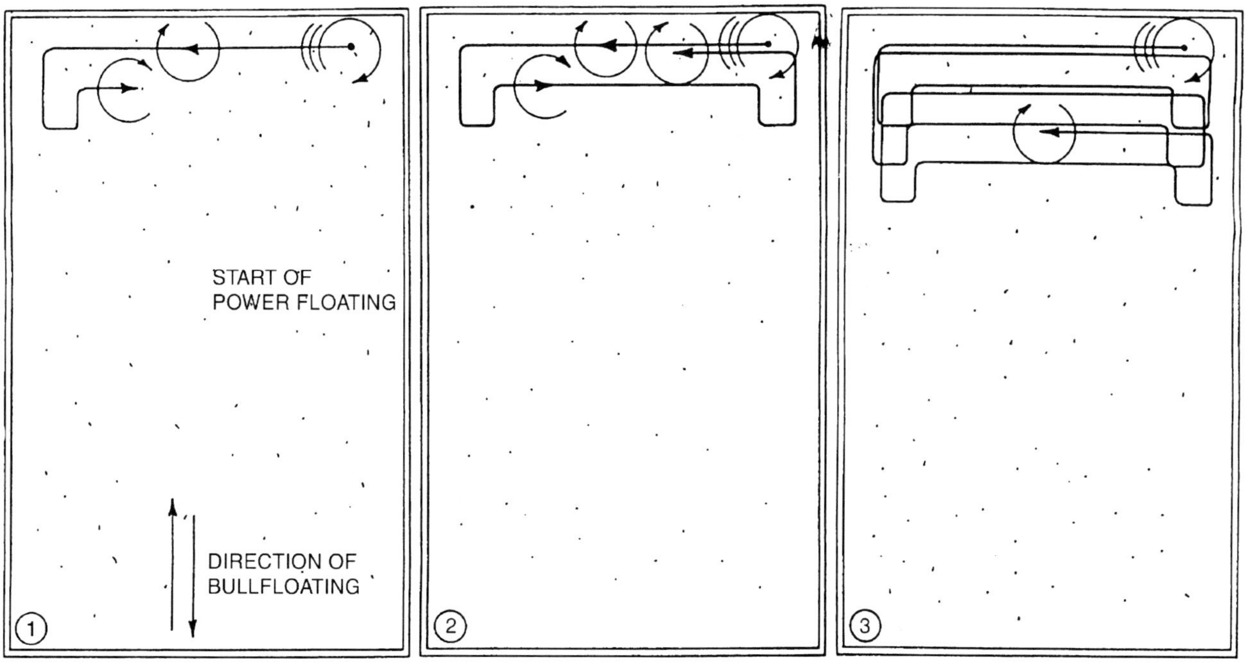

107F16.TIF

TRANSPARENCY 10

TASK MODULE 23107, FINISHING, PART ONE

USING A HAND TROWEL

- Keep pressure on the blade.
- Raise the leading edge slightly for the first pass.
- Move the trowel in an arc.
- Raise the leading edge a little more for each additional pass.

TRANSPARENCY 11

TASK MODULE 23107, FINISHING, PART ONE

FLOATING AND TROWELING

- **Start on knee boards.**
- **Float what you can reach.**
- **Trowel what you floated.**
- **Back yourself away from the finished area.**

TRANSPARENCY 12

TASK MODULE 23107, FINISHING, PART ONE

POWER TROWELING

- Check the equipment and blades.
- Adjust the blades at a slight angle to the surface.
- Use Appropriate Personal Protective Equipment.
- Follow the trowel path.
- Keep the trowel moving.

TRANSPARENCY 13

FIGURE 22

TASK MODULE 23107, FINISHING, PART ONE

TROWEL PATH

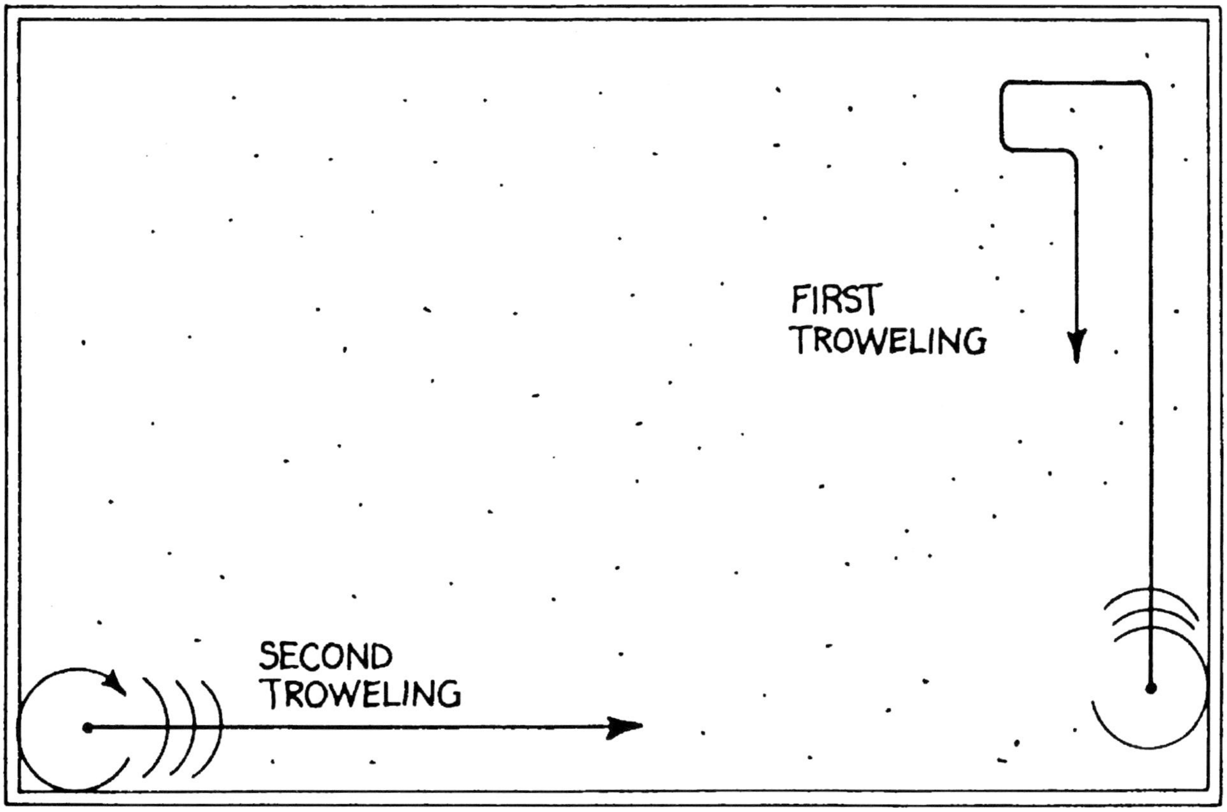

107F23.TIF

TRANSPARENCY 14

FIGURE 23

TASK MODULE 23107, FINISHING, PART ONE

F-NUMBER USE GUIDE

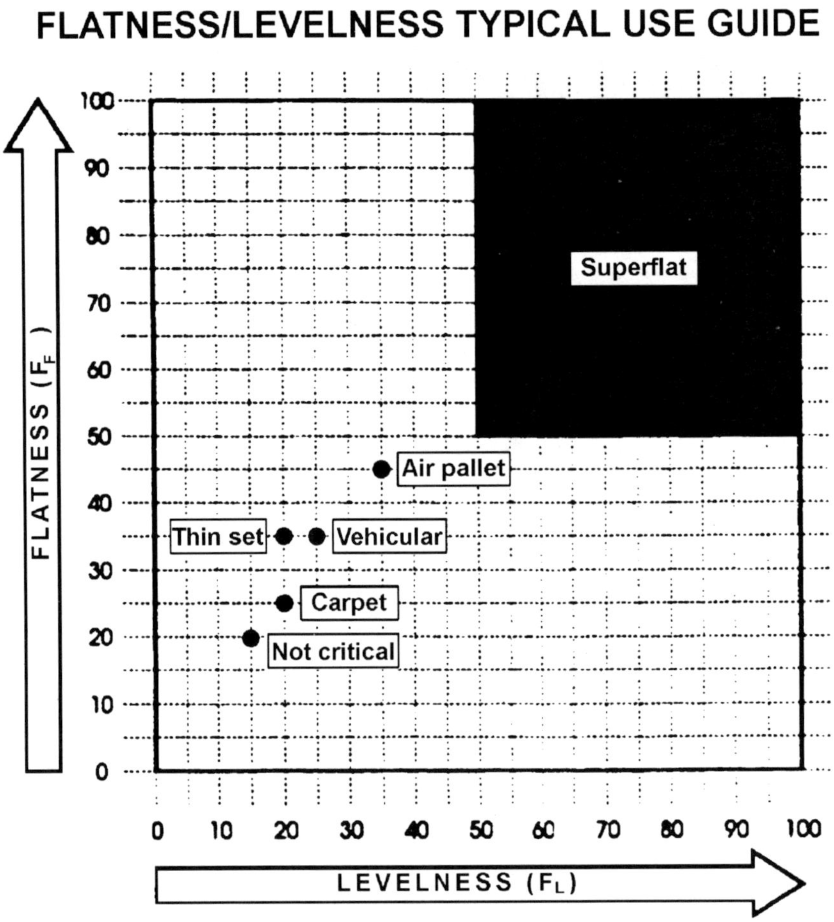

107F23.TIF

TRANSPARENCY 15

FIGURE 26

TASK MODULE 23107, FINISHING, PART ONE

RUBBING

Hand Stone

107F26.TIF

NOTE ON PERFORMANCE PROFILE TESTING

Performance Profiles are included in this Instructor's Guide in a format that can be easily photocopied for each trainee. The Profiles measure trainee competency in the tasks taught in this module.

Please note the number of tasks to be tested while teaching this module. Each trainee should be tested on the tasks listed in the Performance Profile. Before the performance testing, the instructor should brief the trainees on:

- test objectives and criteria,
- safety precautions, and
- procedures for each task to be tested.

The instructor administering the performance testing should also do the following:

- ensure that all of the needed equipment is available and operating properly;
- set up the testing stations;
- organize and administer the test in a way that allows for optimal performance;
- complete the Performance Profile Sheet for each trainee by assigning a score for each listed task;
- monitor adherence to all safety regulations and precautions;
- provide adequate supervision to prevent injuries; and
- take immediate and effective action to remedy any emergency.

ACCREDITATION TESTING

If this Performance Profile Testing is done in the National Center for Construction Education and Research Standardized Craft Training Program, the following conditions must be met:

1. The Craft Instructor must hold valid NCCER instructor certification.
2. The training must be delivered through a Training Program Sponsor recognized by the NCCER.
3. For every module, the specific performance testing must be completed to the satisfaction of the instructor.
4. The results of the testing must be recorded on the Craft Training Report Form. This form must be provided to the local Training Program Sponsor to be forwarded to the NCCER National Craft Training Registry.

Craft: Concrete Finishing

Task Module Number: 23107

Task Module Title: Finishing, Part One

WHEELS OF LEARNING

TRAINEE NAME: ______________________________

TRAINEE SOCIAL SECURITY NUMBER: ______________________________

CLASS: ______________________________

TRAINING PROGRAM SPONSOR: ______________________________

____

INSTRUCTOR: ______________________________

Rating Levels:
1. Passed: performed task.
2. Failed: did not perform task.

Recognition: When testing for the NCCER Standardized Craft Training Program, be sure to record Performance Profile testing results on Craft Training Report Form 200 and submit the results to the Training Program Sponsor.

Objective	TASK	RATING
1, 2, 6	1. Hand float, edge, groove, and trowel a 4 × 4 foot concrete slab.	
1, 4, 6	2. Apply a broom finish to a slab.	
1, 3	3. Mark joints and cut them with a power saw.	
1, 5, 6	4. Apply a rubbing finish to a slab.	

NCCER CRAFT TRAINING USER UPDATES

The NCCER makes every effort to keep these manuals up-to-date and free of technical errors. We appreciate your help in this process. If you have an idea for improving this manual, or if you find an error, a typographical mistake, or an inaccuracy in the NCCER's Craft Training Manuals, please write us, using this form or a photocopy. Be sure to include the exact module number, page number, a description of the problem, and the correction, if possible. Your input will be brought to the attention of the Technical Review Committee. Thank you for your assistance.

Instructors – If you found that additional materials were necessary in order to teach this module effectively, please let us know so that we may include them in the Equipment/Materials list in the Instructor's Guide.

Write: Curriculum and Revision Department
National Center for Construction Education and Research
P.O. Box 141104
Gainesville, FL 32614-1104

Fax: 352-334-0932

Craft ______________________ Module Name ______________________

Module Number ______________________ Page Number(s) ______________________

Description of Problem ______________________

____

____

____

(Optional) Correction of Problem ______________________

____

____

(Optional) Your Name and Address ______________________

____

____

Curing and Protecting Concrete

Module 23108

CURING AND PROTECTING CONCRETE

NATIONAL
CENTER FOR
CONSTRUCTION
EDUCATION AND
RESEARCH

OBJECTIVES

Upon completion of this module, the trainee will be able to:

1. Describe the process of curing concrete.
2. Identify methods of curing concrete.
3. Describe how each method is applied.
4. Identify when each method is used.
5. Associate trade terms with the appropriate processes and equipment.

Note to the Instructor

Before teaching this module, you should review the details in this Instructor's Guide for Equipment and Supplies, Testing, and the suggested Teaching Sequence. Be sure to allow ample time to prepare your own training plan or lesson plan and to gather all required equipment and materials.

Required Equipment and Materials

The following are required for instruction using this module:

Equipment
Overhead projector and screen
Whiteboard/chalkboard
Appropriate Personal Protective Equipment
Manual compound sprayers

Materials
Trainee Task Module
Transparencies
Pencils and paper
Markers/chalk
Water
Plastic sheeting
Scrap lumber and rocks
Wide plastic tape
Samples of curing compounds
Module Examinations
Performance Profile Sheets

HOW TO USE THIS INSTRUCTOR'S GUIDE

For each 2½ hour class session in this Instructor's Guide, the basic Presentation Sequence is as follows:

Introduction/Overview

Classroom, and/or Demonstration, and/or Laboratory

Class Break

Classroom, and/or Demonstration, and/or Laboratory

Summary

Suggested time periods for classroom sessions are included throughout this Instructor's Guide. These time periods should be adapted to meet local conditions and training requirements.

Each class session is presented with two columns of information. On the left side of the page, a narrow column provides suggested trainee and instructor actions, icons to call your attention to material, safety, audiovisual, or testing requirements, and space for your notes. The right-hand column provides the outline of the suggested presentation for each class session.

In this Instructor's Guide, the terms classroom, demonstration, and laboratory are defined and used as follows.

Classroom: Sessions are designed for lectures, group discussions, coaching, and additional activities. Trainees should be encouraged to actively participate.

Demonstration: Instructors will demonstrate all procedures before trainees attempt them. Instructors should make sure that trainees can point out all safety procedures during demonstrations to be assured of the proper use of equipment by trainees.

Laboratory: Instructors will facilitate all laboratory activities, coach trainees as they practice the procedures, monitor trainee progress, and provide feedback. The instructor will make sure that safety rules are followed at all times and that protective equipment is worn.

NCCER Standardized Craft Training Programs

The National Center for Construction Education and Research (NCCER) provides a standardized national program of accredited craft training. Key features of the program include instructor certification, competency-based training, and performance testing. The program provides trainees, instructors, and companies with a standard form of recognition through a National Craft Training Registry. The program is described in full in the *Guidelines for Accreditation*, published by the NCCER. For more information on standardized craft training, contact the NCCER at P.O. Box 141104, Gainesville, FL 32614-1104; or call 352-334-0911.

MODULE OVERVIEW

This module introduces the Concrete Finishing trainee to the methods and procedures used in curing and protecting concrete.

Prerequisites

Please see the Course Map. Prior to training with this module, it is recommended that the trainee shall have successfully completed the following modules:

Core Curricula

Concrete Finishing Level 1, Modules 23101 through 23107

Safety Considerations

Ensure that the trainees are equipped with Appropriate Personal Protective Equipment.

Teaching Time for This Module

Approximately 5 hours or two sessions of training time are suggested to cover *Curing And Protecting Concrete*. The training class session is a suggested 2½ hour time period, which includes at least one break. **You will need to adjust the time required for hands-on activities and testing based on your class size and resources.** All time periods for this module are suggested, and you will need to adapt the suggested lesson plan to meet your local conditions.

Suggested Teaching Sequence — Two 2½-Hour Sessions

Adjust your class times based on class size and resources.

Session	Topic	Trainee Module Section(s)
1	Introduction; Importance of Curing; Curing Methods; Protection	1.0.0 – 4.5.0
2	Laboratory Test Module Examination Performance Profile Test	3.0.0 – 4.5.0

Optional References for Advanced Study

This module is intended to present thorough resources for task training. The following reference works are suggested for both instructors and motivated trainees interested in further study. These are optional materials for continued education rather than for task training.

Concrete Construction and Estimating, Craig Avery, Craftsman Book Company, Carlsbad, CA, 1991.

Concrete Manual, International Conference of Building Officials, Whittier, CA, 1992.

Concrete Parking Lots: Eight Steps to Quality Construction, Videotape, Portland Cement Association, Skokie, IL.

Concrete Technology, Third Edition, George R. White, Delmar Publishers, Inc., Albany, NY, 1991.

Design and Control of Concrete Mixtures, Thirteenth Edition, Portland Cement Association, Skokie, IL, 1988.

Finishing Concrete Flatwork, Videotape 32:00 minutes, Portland Cement Association, Skokie, IL, 1991.

Guide for Concrete Floor and Slab Construction, ACI 302.1R-96, American Concrete Institute, Farmington Hills, MI, 1997.

The Contractor's Guide to Quality Construction, American Concrete Institute, Farmington Hills, MI, 1997.

COURSE MAP

This course map shows all of the modules in the first level of the Concrete Finishing curricula. The suggested training order begins at the bottom and proceeds up. Skill levels increase as a trainee advances on the course map. The training order may be adjusted by the local Training Program Sponsor.

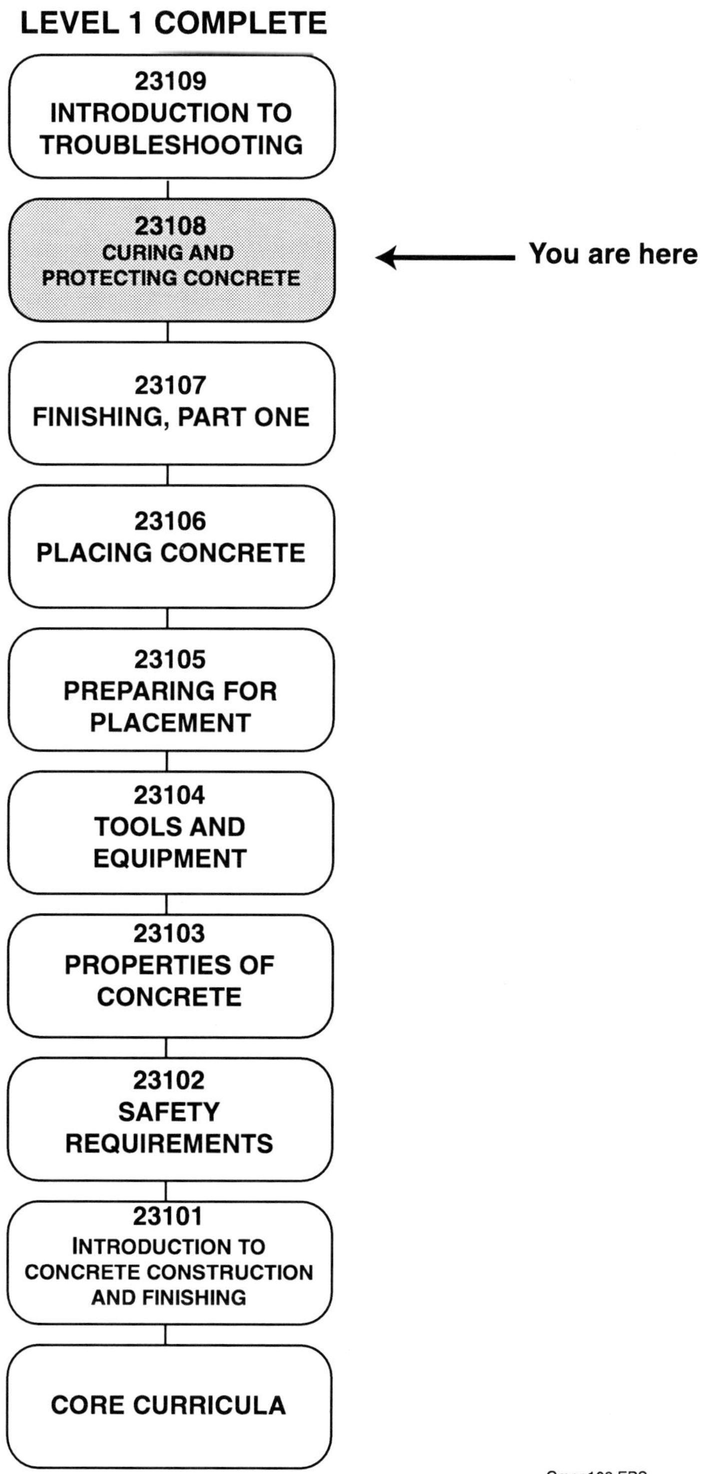

Cmap108.EPS

SESSION 1 OF 2 SESSIONS

Equipment and Materials

Trainee Task Module
Overhead projector and screen
Whiteboard/chalkboard
Markers/chalk
Pencils and paper
Transparencies
Appropriate Personal Protective Equipment
Manual compound sprayers
Water
Plastic sheeting
Wide plastic tape
Scrap lumber and rocks
Samples of curing compounds

Ensure that all trainees have brought their Trainee Modules to class.

Time Schedule (150 minutes)

10 Overview
50 Classroom
10 Class Break
30 Demonstration
5 Class Break
35 Classroom
10 Summary

Overview (10 minutes)

1. Take attendance and perform other administrative tasks.
2. Outline briefly the topics to be covered. Explain to the trainees that they must pass the Module Examination with a grade of at least 70 percent. Also, explain that they must pass the Performance Profile Examination by performing each task to the level of the acceptance criteria.
3. Provide a brief explanation of the acceptance criteria that will be used in the Performance Profile Examination.

Show Transparency 1, Course Objectives.

Show Transparency 2, Performance Profile Tasks.

Classroom (50 minutes)

Use the whiteboard to define and illustrate the Trade Terms.

1. Discuss the trade terms introduced in this module.

Show Transparency 3, Hydration.

2. Review hydration as a chemical reaction between cement and water.

AUDIO/VISUAL

3. Review curing as controlling:
   - Moisture
   - Temperature

Show Transparency 4, The Effect Of Moist-Curing On Concrete Strength With Age (Figure 1, Trainee Module).

Show Transparency 5, Concrete Strength Gain Changes With Temperature (Figure 2, Trainee Module).

4. Discuss how hydration continues over time.
   - Discuss the effects of moisture on strength gain.
   - Discuss the effects of temperature on strength gain.
   - Review the effects of hot weather on concrete curing.
   - Review the effects of cold weather on concrete curing.
   - Review the effects of curing on slab surfaces.

Show Transparencies 6 and 7, Curing Methods, Advantages, and Limitations (Figure 3, Trainee Module).

5. Discuss the three main methods of curing:
   - Adding water to the surface
   - Preventing water loss from the surface
   - Adding heat and/or moisture
6. Ask trainees to identify each of the types of curing in *Figure 3* in the Trainee Task Module according to method.

Show Transparency 8, Sprinkling System (Figure 4, Trainee Module).

7. Discuss the advantages of curing by sprinkling.

Show Transparency 9, Rewetting A Wet Covering (Figure 5, Trainee Module).

8. Discuss the advantages of curing by wet covering.
9. Describe the various types of wet coverings used in your area.

Show Transparency 10, Placing Waterproof Paper (Figure 6, Trainee Module).

10. Discuss the advantages of curing by impervious paper.

Show Transparency 11, Surface Discoloration Under Plastic (Figure 8, Trainee Module).

11. Discuss the advantages of curing by plastic sheeting. Emphasize the importance of:
    - Sealing edges
    - Removing wrinkles
12. Discuss the advantages of curing by membrane-forming compounds. Emphasize the importance of:
    - Covering all exposed surfaces, including edges
    - Spraying in a uniform pattern

Show Transparency 12, Compound Spraying Pattern (Figure 10, Trainee Module).

Pass around containers with samples of different types of curing compounds.

13. Discuss different types of curing compounds:
    - Clear wax-resin base
    - Clear resin
    - Black and gray-pigmented asphalt
    - New chlorinated rubber, acrylics, and epoxies

Class Break (10 minutes)

Demonstration (30 minutes)

1. Mark out slab area on floor with plastic tape or scrap lumber.
2. Give the operating procedure and a short demonstration of sprayer functions.
3. Discuss and demonstrate the appropriate safety procedures.
4. Demonstrate how to apply a membrane-forming compound to a slab.
   - Demonstrate spraying pattern shown in *Figure 10.*
5. Demonstrate to trainees how to apply plastic sheeting to a wet surface.
   - Demonstrate removing wrinkles.
   - Demonstrate sealing edges with tape or lumber.
6. Review procedures and answer trainee questions.

Class Break (5 minutes)

Classroom (35 minutes)

Show Transparency 13, Hand-Made Insulating Blanket (Figure 12, Trainee Module).

1. Discuss the advantages of curing with insulating blankets.
2. Discuss the advantages of curing by keeping forms on.

Show Transparency 14, Live Steam Cycle (Figure 14, Trainee Module).

3. Discuss steam curing and other heat curing.

Show Transparency 5, Concrete Strength Changes With Temperature (Figure 2, Trainee Module).

4. Discuss hot weather problems in curing:
   - High internal temperatures
   - Rapid evaporation of water

Show Transparencies 15 and 16, Hot Weather Techniques.

5. Explain different hot weather curing techniques in text.
   - Ask trainees to identify which problems each technique addresses.
6. Discuss cold weather problems in curing:
   - Low internal temperatures
   - Freezing

Show Transparencies 17 and 18, Cold Weather Techniques.

7. Go through bulleted list of cold weather curing techniques.
   - Ask trainees to identify which problems each technique addresses.

8. Discuss the advantages of temporary winter enclosures.
9. Discuss the techniques for dealing with the effects of wind on fresh slab surfaces.
10. Discuss the techniques for dealing with the effects of rain on fresh slab surfaces.
11. Discuss the techniques for dealing with the effects of traffic over fresh slab surfaces.

Summary (10 minutes)

1. Review the materials covered in this session.
2. Answer trainee questions.
3. Give assignments for the next session. Have trainees complete the review/practice questions in the Trainee Module before the next session.
4. Explain that the next session will be the last session and includes the Module Examination and the Performance Profile Examination.

SESSION 2 OF 2 SESSIONS

Ensure that all trainees have brought their Trainee Modules to class.

Equipment and Materials

Trainee Module
Overhead projector and screen
Whiteboard/chalkboard
Markers/chalk
Manual compound sprayers
Water
Plastic sheeting
Wide plastic tape
Scrap lumber and rocks
Appropriate Personal Protective Equipment
Module Examinations
Performance Profile Sheets

Time Schedule (150 minutes)

5 Overview
55 Classroom/Laboratory
10 Class Break
30 Module Examination
10 Class Break
40 Performance Profile Examination

Overview (5 minutes)

1. Take attendance and perform other administrative tasks.
2. Outline briefly the topics to be covered. Explain to the trainees that they must pass the Module Examination with a grade of at least 70 percent. Also, explain that they must pass the Performance Profile Examination by performing each task to the level of acceptance criteria.
3. Provide a brief explanation of the acceptance criteria that will be used in the Performance Profile Examination.

Classroom/Laboratory (55 minutes)

1. Mark slab areas on floor with tape.
2. Assign equipment and tasks to trainees.
3. Review safety procedures and sprayer operation procedures.
4. Coach and provide feedback as trainees practice spraying in a patterned spray.

5. Coach and provide feedback as trainees practice laying plastic sheeting on a wet surface.
6. Review procedures and answer trainee questions.
7. Discuss review/practice questions and address any questions or comments from trainees.
8. Review key concepts of this module.

Class Break (10 minutes)

Module Examination (30 minutes)

1. Hand out Module Examinations to all trainees. Tell them they have 30 minutes to complete the exam.
2. Collect and grade the examinations.

Class Break (10 minutes)

Performance Profile Examination (40 minutes)

1. Have trainees perform the following performance objectives to the satisfaction of the instructor:
   - Spray a compound in a systematic total-coverage pattern over a marked slab surface
   - Cover a marked wet slab surface with plastic sheeting, sealed and wrinkle free
2. Fill out Performance Profile Sheets for each trainee.

MODULE EXAMINATIONS AND ANSWER KEYS

The masters for Module Examinations and Answer Keys are found in the booklet entitled "Concrete Finishing Level One Module Examinations" included in this Instructor's Guide package. You may also reproduce the tests using the Windows-compatible software program which may be purchased separately from Prentice Hall. If you do not have the reproducible masters in your Instructor's Guide package, or would like to order the test software, please contact Prentice Hall Publishers at 1-800-922-0579.

Be sure to keep answer keys and tests in a safe area.

OVERHEAD TRANSPARENCIES

Masters for overhead transparencies are found on the following pages. Transparencies can be made by using clear acetate sheets in a photocopy machine. This can be done at any commercial copy shop, or by following this method on your office machine:

1. Use standard 8½ × 11 clear acetate sheets, which can be obtained at office supply stores, stationery stores, or commercial copying centers.
2. Check the photocopier's manual to make sure it will take the thicker acetate sheets.
3. Remove the paper supply in the photocopying machine and replace it with a supply of acetate sheets.
4. Make photocopies from the masters on the following pages in the usual way.
   The material will be reproduced on the clear acetate sheets for use on any overhead projector.

TRANSPARENCY 1

TASK MODULE 23108, CURING AND PROTECTING CONCRETE

COURSE OBJECTIVES

Upon completion of this module, the trainee will be able to:

1. **Describe the process of curing concrete.**
2. **Identify methods of curing concrete.**
3. **Describe how each method is applied.**
4. **Identify when each method is used.**
5. **Associate trade terms with the appropriate processes and equipment.**

TRANSPARENCY 2

TASK MODULE 23108, CURING AND PROTECTING CONCRETE

PERFORMANCE PROFILE TASKS

1. Spray a compound in a systematic total-coverage pattern over a marked slab surface.

2. Cover a marked wet slab surface with plastic sheeting, sealed and wrinkle free.

TRANSPARENCY 3

TASK MODULE 23108, CURING AND PROTECTING CONCRETE

HYDRATION

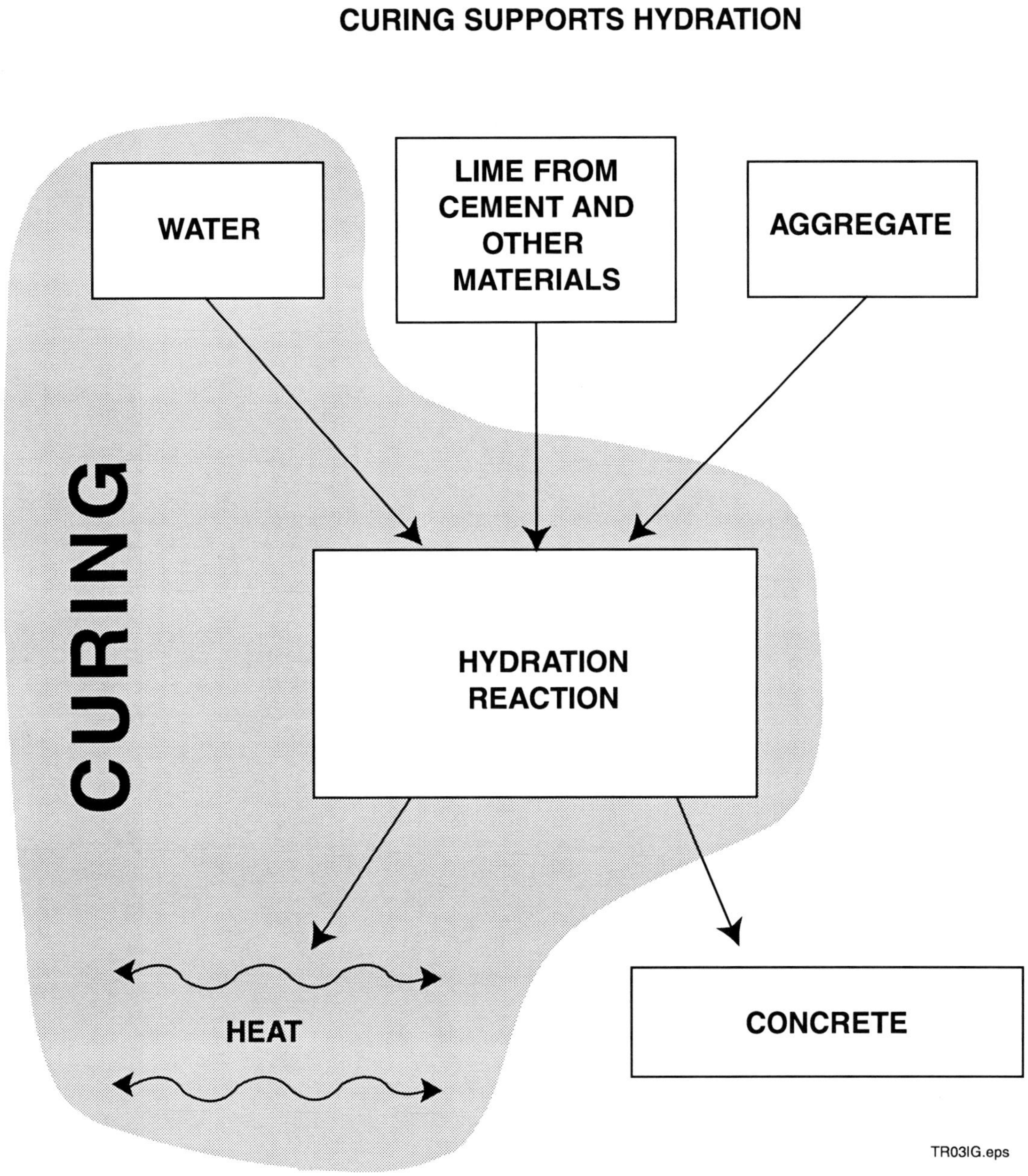

TRANSPARENCY 4

FIGURE 1

TASK MODULE 23108, CURING AND PROTECTING CONCRETE

THE EFFECT OF MOIST-CURING ON CONCRETE STRENGTH WITH AGE

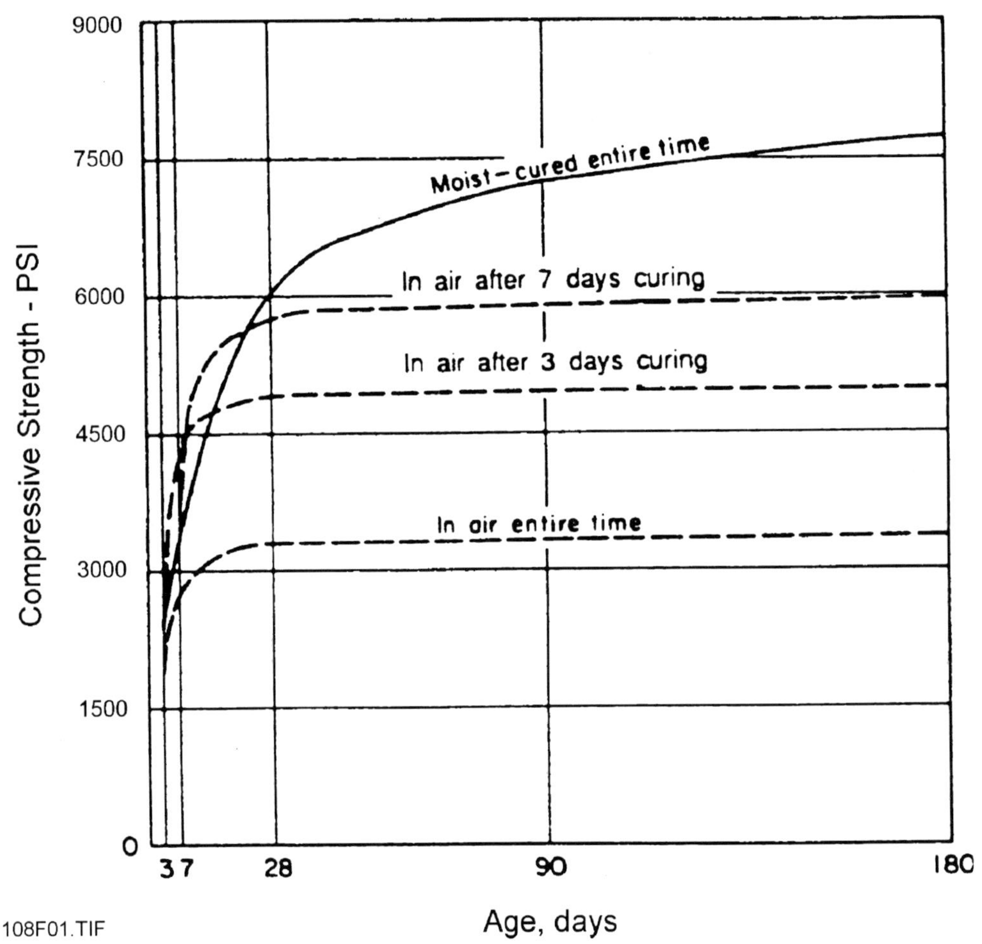

108F01.TIF

TRANSPARENCY 5

FIGURE 2

TASK MODULE 23108, CURING AND PROTECTING CONCRETE

CONCRETE STRENGTH GAIN CHANGES WITH TEMPERATURE

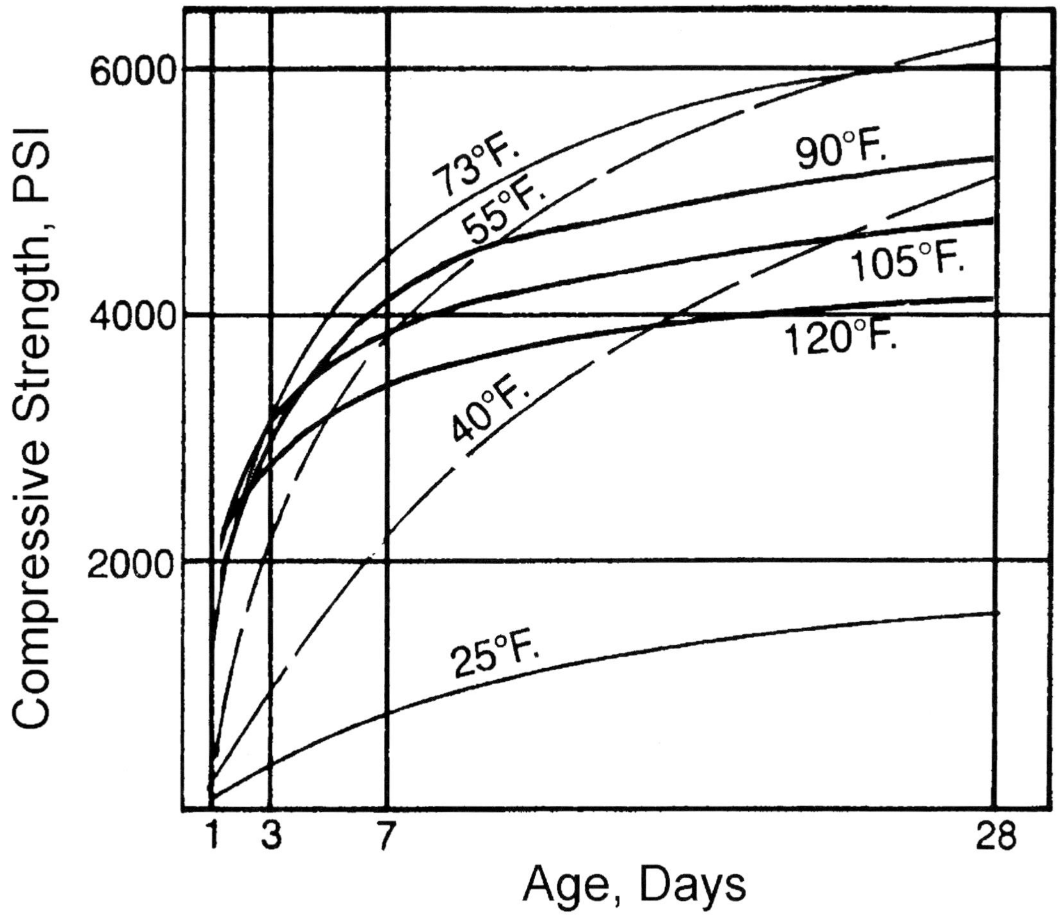

108F02.TIF

TRANSPARENCY 6

FIGURE 3

TASK MODULE 23108, CURING AND PROTECTING CONCRETE

CURING METHODS, ADVANTAGES, AND LIMITATIONS

Type of Curing	Where to Use	Advantages	Limitations
Spraying/ fogging	Horizontal or vertical work (slabs or walls)	Thorough moist curing, very clean surface for bonding	Needs water source, relies on equipment (timers) or workers to spray at appropriate intervals, drying out can occur, prevents other work from occurring on slab, affected by wind
Wet coverings	Horizontal or vertical work (slabs or walls)	Thorough moist curing, very clean surface for bonding, can help control temperature	Coverings may be messy or difficult to use/remove, must be continually wetted, drying out of covering can draw moisture away from slab, prevents other work from occurring on slab, can discolor surface, potential fire hazard
Impervious paper	Horizontal work (slabs)	Very clean surface for bonding, allows using slab as work surface, inexpensive, will not stain	Must be anchored and sealed at edges and joints

TR06IG.EPS

TRANSPARENCY 7

FIGURE 3

TASK MODULE 23108, CURING AND PROTECTING CONCRETE

CURING METHODS , ADVANTAGES, AND LIMITATIONS (Cont.)

Plastic sheeting	Horizontal work (slabs)	Very clean surface for bonding, allows using slab as work surface, inexpensive, can help control temperature, reusable	Must be anchored and sealed at edges and joints, may discolor concrete, can tear under traffic, not for use where appearance is important
Membrane-forming compounds	Horizontal or vertical work (slabs or walls)	Fast and easy application, inexpensive, can help control surface temperature, may allow traffic on horizontal work after short period	Can interfere with surface bondings of other materials (paints, coatings, tile), may contain solvents
Insulating blankets	Anywhere (slabs, walls, or other formed surfaces)	Retains both heat and moisture, flexible	Must be shaped to member and anchored or held in place
Live steam	Anywhere (slabs, walls, or other formed surfaces)	Supplies heat and moisture, increases the rate of strength gain	Requires enclosed area, requires special equipment, time-consuming setup, expensive, can be dangerous

TR07IG.EPS

TRANSPARENCY 8

FIGURE 4

TASK MODULE 23108, CURING AND PROTECTING CONCRETE

SPRINKLING SYSTEM

108F04.TIF

TRANSPARENCY 9

FIGURE 5

TASK MODULE 23108, CURING AND PROTECTING CONCRETE

REWETTING A WET COVERING

108F05.TIF

TRANSPARENCY 10

FIGURE 6

TASK MODULE 23108, CURING AND PROTECTING CONCRETE

PLACING WATERPROOF PAPER

108F06.TIF

TRANSPARENCY 11

FIGURE 8

TASK MODULE 23108, CURING AND PROTECTING CONCRETE

SURFACE DISCOLORATION UNDER PLASTIC

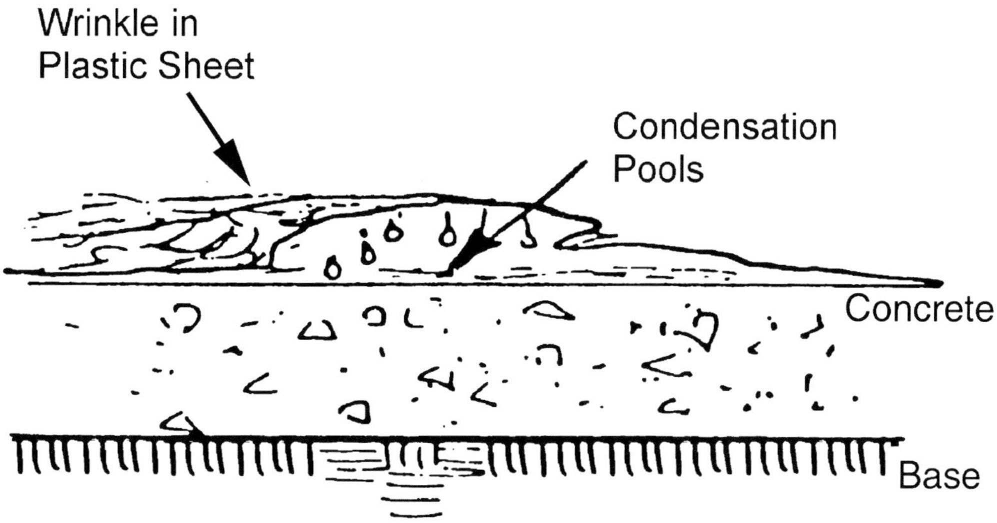

108F08.TIF

TRANSPARENCY 12

FIGURE 10

TASK MODULE 23108, CURING AND PROTECTING CONCRETE

COMPOUND SPRAYING PATTERN

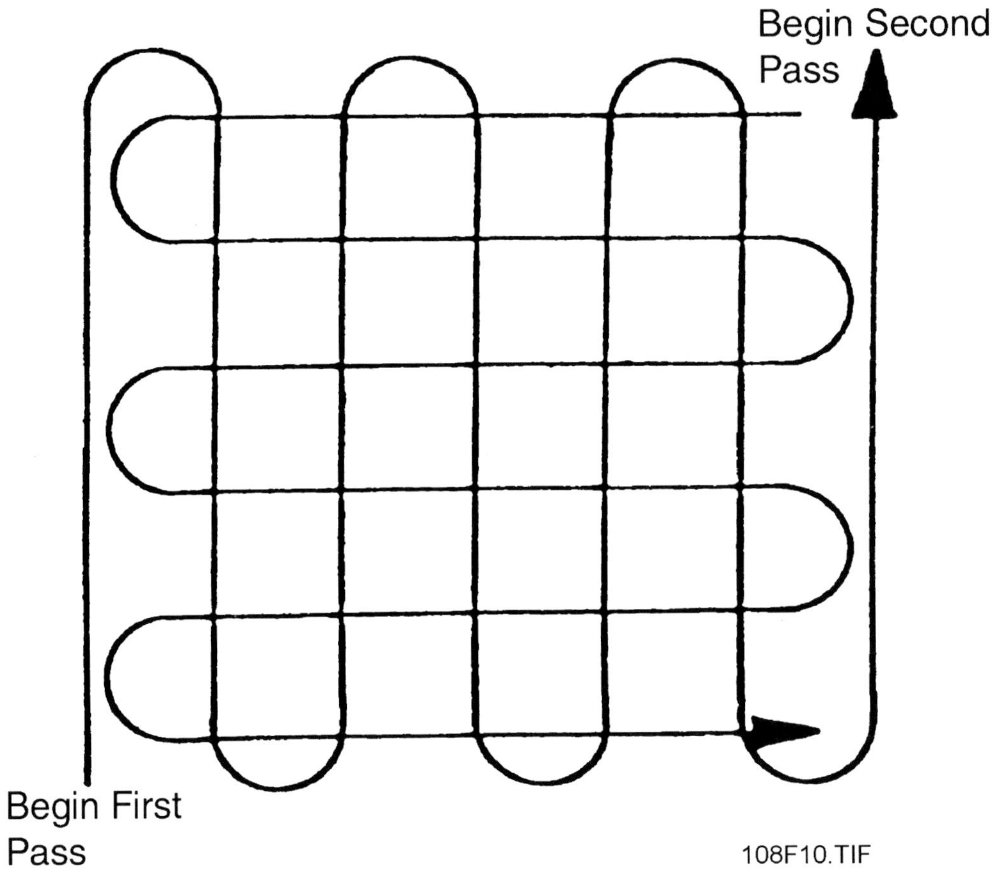

TRANSPARENCY 13

FIGURE 12

TASK MODULE 23108, CURING AND PROTECTING CONCRETE

HAND-MADE INSULATING BLANKET

108F12.TIF

TRANSPARENCY 14

FIGURE 14

TASK MODULE 23108, CURING AND PROTECTING CONCRETE

LIVE STEAM CYCLE

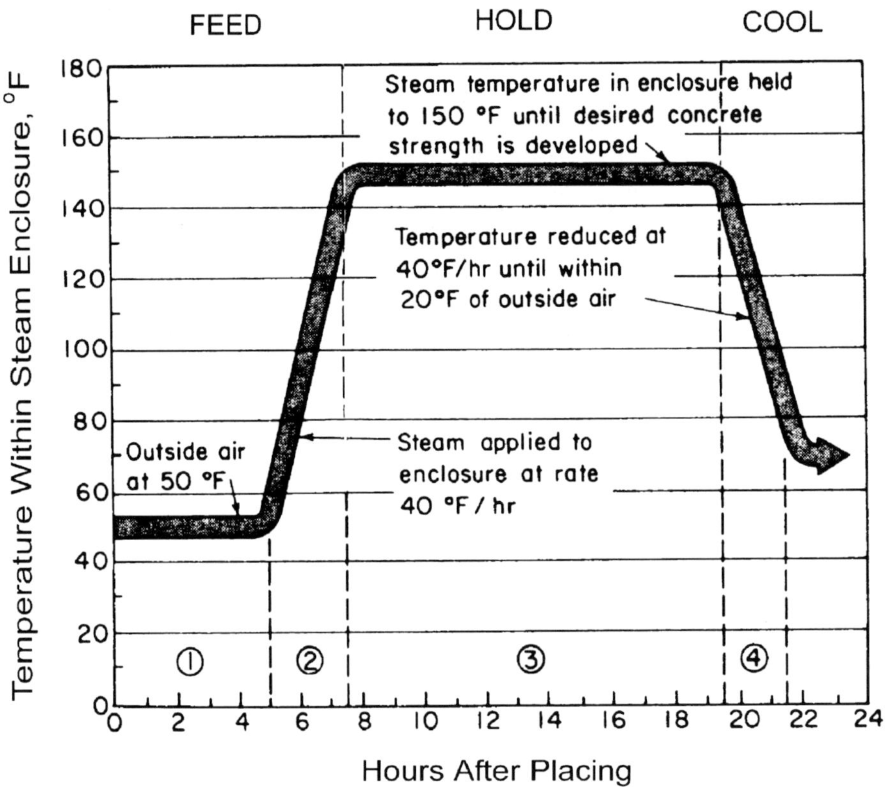

① Initial delay prior to steaming 2 to 5 hours
② Temperature increase period 2 1/2 hours
③ Constant temperature period 6 to 12 hours or more
④ Temperature decrease period 2 hours

108F14.TIF

TRANSPARENCY 15

TASK MODULE 23108, CURING AND PROTECTING CONCRETE

HOT WEATHER TECHNIQUES

- Monitor fresh concrete temperature.
- Work and schedule work rapidly.
- Cool mix ingredients, especially water. Use ice as part of the mix water, or add shaved ice directly to the mixer.
- Use a retarding admixture.
- Protect curing concrete by paying special attention to retaining moisture.
- Be sure the subgrade is thoroughly moistened in advance.
- Wet the forms in advance, or use a release agent on them.

TRANSPARENCY 16

TASK MODULE 23108, CURING AND PROTECTING CONCRETE

HOT WEATHER TECHNIQUES (Cont.)

- **Use continuous spraying.**
- **Use white-pigmented compounds, plastic sheeting, or wet coverings.**
- **Taper off curing gradually.**
- **Place concrete during cooler parts of the day.**
- **Use mineral admixtures to reduce the heat of hydration.**

TRANSPARENCY 17

TASK MODULE 23108, CURING AND PROTECTING CONCRETE

COLD WEATHER TECHNIQUES

- **Prepare in advance with heaters on hand.**
- **Heat water and materials added to concrete.**
- **Use an accelerating admixture.**
- **Use air-entrained concrete.**
- **Use insulating blankets overnight.**
- **Do not place concrete on a frozen subgrade.**
- **Preheat form surfaces.**
- **Wrap and seal insulating blankets around forms.**

TRANSPARENCY 18

TASK MODULE 23108, CURING AND PROTECTING CONCRETE

COLD WEATHER TECHNIQUES (Cont.)

- Avoid placing concrete during extreme cold weather.
- Use black-pigmented compounds, black plastic sheeting, or kraft paper.
- Taper off curing gradually, so concrete is not shocked by sudden cold.
- Monitor fresh concrete temperatures.

NOTE ON PERFORMANCE PROFILE TESTING

Performance Profiles are included in this Instructor's Guide in a format that can be easily photocopied for each trainee. The Profiles measure trainee competency in the tasks taught in this module.

Please note the number of tasks to be tested while teaching this module. Each trainee should be tested on the tasks listed in the Performance Profile. Before the performance testing, the instructor should brief the trainees on:

- test objectives and criteria,
- safety precautions, and
- procedures for each task to be tested.

The instructor administering the performance testing should also do the following:

- ensure that all of the needed equipment is available and operating properly;
- set up the testing stations;
- organize and administer the test in a way that allows for optimal performance;
- complete the Performance Profile Sheet for each trainee by assigning a score for each listed task;
- monitor adherence to all safety regulations and precautions;
- provide adequate supervision to prevent injuries; and
- take immediate and effective action to remedy any emergency.

ACCREDITATION TESTING

If this Performance Profile Testing is done in the National Center for Construction Education and Research Standardized Craft Training Program, the following conditions must be met:

1. The Craft Instructor must hold valid NCCER instructor certification.
2. The training must be delivered through a Training Program Sponsor recognized by the NCCER.
3. For every module, the specific performance testing must be completed to the satisfaction of the instructor.
4. The results of the testing must be recorded on the Craft Training Report Form. This form must be provided to the local Training Program Sponsor to be forwarded to the NCCER National Craft Training Registry.

Craft: Concrete Finishing	WHEELS
Task Module Number: 23108	OF
Task Module Title: Curing And Protecting Concrete	LEARNING

TRAINEE NAME: ______________________________

TRAINEE SOCIAL SECURITY NUMBER: ______________________________

CLASS: ______________________________

TRAINING PROGRAM SPONSOR: ______________________________

____

INSTRUCTOR: ______________________________

Rating Levels:
1. Passed: performed task.
2. Failed: did not perform task.

Recognition: When testing for the NCCER Standardized Craft Training Program, be sure to record Performance Profile testing results on Craft Training Report Form 200 and submit the results to the Training Program Sponsor.

Objective	TASK	RATING
1	1. Spray a compound in a systematic total-coverage pattern over a marked slab surface.	
2	2. Cover a marked wet slab surface with plastic sheeting, sealed and wrinkle free.	

NCCER CRAFT TRAINING USER UPDATES

The NCCER makes every effort to keep these manuals up-to-date and free of technical errors. We appreciate your help in this process. If you have an idea for improving this manual, or if you find an error, a typographical mistake, or an inaccuracy in the NCCER's Craft Training Manuals, please write us, using this form or a photocopy. Be sure to include the exact module number, page number, a description of the problem, and the correction, if possible. Your input will be brought to the attention of the Technical Review Committee. Thank you for your assistance.

Instructors – If you found that additional materials were necessary in order to teach this module effectively, please let us know so that we may include them in the Equipment/Materials list in the Instructor's Guide.

Write: Curriculum and Revision Department
National Center for Construction Education and Research
P.O. Box 141104
Gainesville, FL 32614-1104

Fax: 352-334-0932

Craft ______________________ Module Name ______________________

Module Number ______________________ Page Number(s) ______________________

Description of Problem ______________________

____

____

____

(Optional) Correction of Problem ______________________

____

____

(Optional) Your Name and Address ______________________

____

____

notes

Introduction to Troubleshooting

Module 23109

Concrete Finishing Instructor's Guide Task Module 23109

INTRODUCTION TO TROUBLESHOOTING

NATIONAL
CENTER FOR
CONSTRUCTION
EDUCATION AND
RESEARCH

OBJECTIVES

Upon completion of this module, the trainee will be able to:

1. Describe a basic troubleshooting methodology that can be used to identify a variety of concrete construction problems and their causes.
2. Identify problems with fresh concrete and describe ways to prevent them.
3. Identify different concrete defects such as crazing, cracking, dusting, scaling, popouts, and efflorescence, and describe ways to prevent them.

Note to the Instructor

Before teaching this module, you should review the details in this Instructor's Guide for Equipment and Supplies, Testing, and the suggested Teaching Sequence. Be sure to allow ample time to prepare your own training plan or lesson plan and to gather all required equipment and materials.

Required Equipment and Materials

The following are required for instruction using this module:

Equipment
Overhead projector and screen
Whiteboard/chalkboard
Appropriate Personal Protective Equipment

Materials
Trainee Task Module
Transparencies
Markers/chalk
Paper and pencils
Photographs of concrete problems
Module Examinations
Performance Profile Sheets

HOW TO USE THIS INSTRUCTOR'S GUIDE

For each 2½ hour class session in this Instructor's Guide, the basic Presentation Sequence is as follows:

Introduction/Overview

Classroom, and/or Demonstration, and/or Laboratory

Class Break

Classroom, and/or Demonstration, and/or Laboratory

Summary

Suggested time periods for classroom sessions are included throughout this Instructor's Guide. These time periods should be adapted to meet local conditions and training requirements.

Each class session is presented with two columns of information. On the left side of the page, a narrow column provides suggested trainee and instructor actions, icons to call your attention to material, safety, audiovisual, or testing requirements, and space for your notes. The right-hand column provides the outline of the suggested presentation for each class session.

In this Instructor's Guide, the terms classroom, demonstration, and laboratory are defined and used as follows:

Classroom: Sessions are designed for lectures, group discussions, coaching, and additional activities. Trainees should be encouraged to actively participate.

Demonstration: Instructors will demonstrate all procedures before trainees attempt them. Instructors should make sure that trainees can point out all safety procedures during demonstrations to be assured of the proper use of equipment by trainees.

Laboratory: Instructors will facilitate all laboratory activities, coach trainees as they practice the procedures, monitor trainee progress, and provide feedback. The instructor will make sure that safety rules are followed at all times and that protective equipment is worn.

NCCER Standardized Craft Training Programs

The National Center for Construction Education and Research (NCCER) provides a standardized national program of accredited craft training. Key features of the program include instructor certification, competency-based training, and performance testing. The program provides trainees, instructors, and companies with a standard form of recognition through a National Craft Training Registry. The program is described in full in the *Guidelines for Accreditation*, published by the NCCER. For more information on standardized craft training, contact the NCCER at P.O. Box 141104, Gainesville, FL 32614-1104; or call 352-334-0911.

MODULE OVERVIEW

This course introduces the Concrete Finishing trainee to the methods and procedures used in troubleshooting the placement and finishing of fresh and hardened concrete.

Prerequisites

Please see the Course Map. Prior to training with this module, it is recommended that the trainee shall have successfully completed the following modules:

Core Curricula

Concrete Finishing Level 1, Modules 23101 through 23108

Safety Considerations

Ensure that the trainees are equipped with Appropriate Personal Protective Equipment.

Teaching Time for This Module

Approximately 5 hours or two sessions of training time are suggested to cover *Introduction to Troubleshooting*. The training class session is a suggested 2½ hour time period, which includes at least one break. **You will need to adjust the time required for hands-on activities and testing based on your class size and resources.** All time periods for this module are suggested, and you will need to adapt the suggested lesson plan to meet your local conditions.

Suggested Teaching Sequence — Two 2½-Hour Sessions

Adjust your class times based on class size and resources.

Session	Topic	Trainee Module Section(s)
1	Introduction; Troubleshooting Method; Fresh Concrete; Hardened Concrete	1.0.0 – 4.11.0
2	Module Examination; Performance Profile Test	

Optional References for Advanced Study

This module is intended to present thorough resources for task training. The following reference works are suggested for both instructors and motivated trainees interested in further study. These are optional materials for continued education rather than for task training.

Concrete Construction and Estimating, Craig Avery, Craftsman Book Company, Carlsbad, CA, 1991.

Concrete Manual, International Conference of Building Officials, Whittier, CA, 1992.

Concrete Parking Lots: Eight Steps to Quality Construction, Videotape, Portland Cement Association, Skokie, IL.

Concrete Technology, Third Edition, George R. White, Delmar Publishers, Inc., Albany, NY, 1991.

Design and Control of Concrete Mixtures, Thirteenth Edition, Portland Cement Association, Skokie, IL, 1988.

Finishing Concrete Flatwork, Videotape 32:00 minutes, Portland Cement Association, Skokie, IL, 1991.

Guide for Concrete Floor and Slab Construction, ACI 302.1R-96, American Concrete Institute, Farmington Hills, MI, 1997.

The Contractor's Guide to Quality Construction, American Concrete Institute, Farmington Hills, MI, 1997.

COURSE MAP

This course map shows all of the modules in the first level of the Concrete Finishing curricula. The suggested training order begins at the bottom and proceeds up. Skill levels increase as a trainee advances on the course map. The training order may be adjusted by the local Training Program Sponsor.

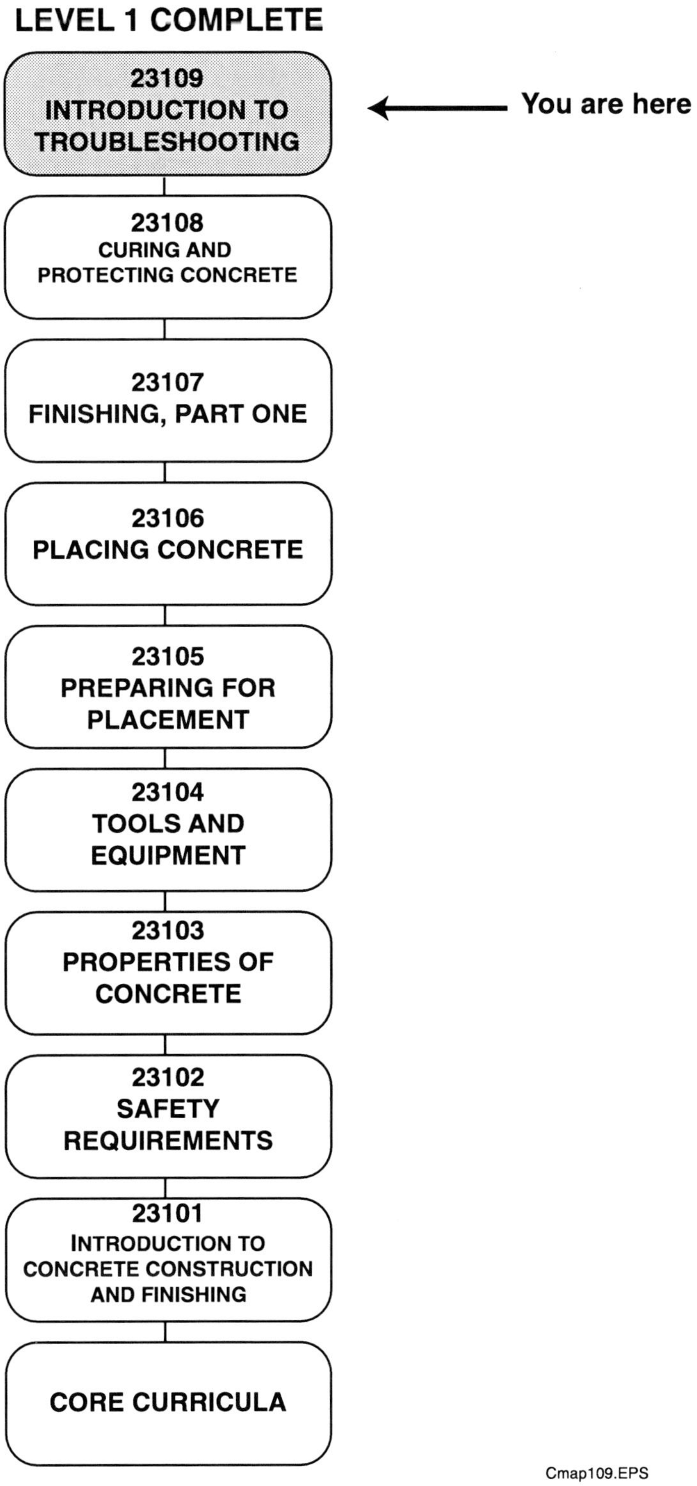

Cmap109.EPS

SESSION 1 OF 2 SESSIONS

Ensure that all trainees have brought their Trainee Modules to class.

Equipment and Supplies

Trainee Task Module
Overhead projector and screen
Whiteboard/chalkboard
Markers/chalk
Transparencies
Pencils and paper
Appropriate Personal Protective Equipment
Photographs of concrete problems

Time Schedule (150 minutes)

10 Overview
60 Classroom
10 Class Break
60 Classroom
10 Summary

Overview (10 minutes)

Show Transparency 1, Course Objectives.

Show Transparency 2, Performance Profile Tasks.

1. Take attendance and perform other administrative tasks.
2. Outline briefly the topics to be covered. Explain to the trainees that they must pass the Module Examination with a grade of at least 70 percent. Also, explain that they must pass the Performance Profile Examination by performing each task to the level of the acceptance criteria.
3. Provide a brief explanation of the acceptance criteria that will be used in the Performance Profile Examination.

Classroom (60 minutes)

1. Review the trade terms introduced in this module and briefly explain them.
2. Explain why it is important to have a good finished product.
   - Structurally sound and safe
   - Reduces costly rework and repairs
   - Provides a good visual appearance
   - Reduces maintenance requirements

Show Transparency 3, Safety Requirements.

3. Point out the need to wear Appropriate Personal Protective Equipment when working with fresh concrete.

Show Transparency 4, Troubleshooting Methodology (Figure 1, Trainee Module).

4. Discuss the three main steps to troubleshooting a problem.
   - Investigate the symptoms
   - Research the problem
   - Apply corrective action
5. Describe how to investigate the symptoms of a problem.
6. Explain the process for researching the symptoms in order to identify the problem.

Show Transparency 5, Corrective Action.

7. Discuss how to select appropriate corrective actions in order to solve a problem. Point out the three requirements the corrective action should meet.

Use whiteboard to list each type of problem and note corrective actions used in your area.

Pass around photographs of the various types of problems if available.

8. Describe the most common problems with fresh concrete.
   - Excess bleeding
   - Segregation
   - Sticky finish
   - Rapid set
   - Slow set
   - Plastic shrinkage cracking
   - Low spots

   Refer to the chart in *Figure 2* of the Trainee Task Module. Discuss specific problems that occur with fresh concrete in your area.

Class Break (10 minutes)

Classroom (60 minutes)

1. Discuss the problems that show up in hardened concrete as a result of problems during placing, finishing, and curing. Point out that nothing can be done to eliminate the problem with the hardened concrete. In this case, only repairs can be made.

   The object here is to inform the trainees of symptoms that appear after the concrete has hardened and to make them aware of the activities that should be performed during placing, finishing, and curing that would eliminate these problems.

Use whiteboard to list each type of problem and note corrective actions used in your area.

Pass around photographs of the various types of problems if available.

2. Describe the most common problems found with hardened concrete.
   - Crazing
   - Shrinkage cracks
   - Dusting
   - Scaling
   - Spalling
   - Popouts
   - Blistering
   - Honeycombs
   - Voids

   Refer to the chart in *Figure 6* of the Trainee Task Module. Discuss specific problems that occur in your area.

Summary (10 minutes)

1. Review the key points for solving problems when placing, finishing, and curing concrete.
2. Answer any questions the trainees may have about problems with fresh or hardened concrete.
3. Tell trainees that the next session will be the last session and includes the Module Examination and the Performance Profile Test.

SESSION 2 OF 2 SESSIONS

Ensure that all trainees have brought their Trainee Modules to class.

Equipment and Supplies

Trainee Task Module
Overhead projector and screen
Whiteboard/chalkboard
Markers/chalk
Pencils and paper
Transparencies
Module Examinations
Performance Profile Sheets

Time Schedule (150 minutes)

10 Overview
70 Classroom
10 Class Break
30 Module Examination
10 Class Break
20 Performance Profile Examination

Show Transparency 1, Course Objectives.

Show Transparency 2, Performance Profile Tasks.

Overview (10 minutes)

1. Take attendance and perform other administrative tasks.
2. Outline briefly the topics to be covered. Explain to the trainees that they must pass the Module Examination with a grade of at least 70 percent. Also, explain that they must pass the performance exercise by performing each task to the point of the acceptance criteria.
3. Provide a brief explanation of the acceptance criteria which will be used in the Performance Profile Examination.

Classroom (70 minutes)

Use whiteboard to list problems and corrective actions to be taken.

1. Review the main problems with finishing fresh concrete in your area, and discuss the corrective steps that need to be taken.
2. Review the main problems identified in hardened concrete. Discuss the correct procedures that need to be followed to keep the problem from recurring.
3. Have trainees complete the Review/Practice Questions in the back of the Trainee Task Module. Point out that the answers to the questions are in the back of the module.
4. Go over the questions and answers with the trainees and explain any problems.

Class Break (10 minutes)

Module Examination (30 minutes)

1. Hand out the Module Examination to all trainees. Tell them they have 30 minutes to complete the exam.
2. Collect and grade the examinations.

Class Break (10 minutes)

Performance Profile Examination (20 minutes)

You can use the set of photographs provided in this module or, if available, your own photographs which may represent local situations more clearly.

1. Have each trainee identify the following types of concrete problems from photographs:
   - Dusting
   - Crazing
   - Spalling
   - Popouts
   - Honeycombs
   - Scaling
   - Plastic shrinkage cracking
   - Excess bleeding
   - Uncontrolled shrinkage cracks
   - Bugholes
2. Fill out Performance Profile Sheets for each trainee.

MODULE EXAMINATIONS AND ANSWER KEYS

The masters for Module Examinations and Answer Keys are found in the booklet entitled "Concrete Finishing Level One Module Examinations" included in this Instructor's Guide package. You may also reproduce the tests using the Windows-compatible software program which may be purchased separately from Prentice Hall. If you do not have the reproducible masters in your Instructor's Guide package, or would like to order the test software, please contact Prentice Hall Publishers at 1-800-922-0579.

Be sure to keep answer keys and tests in a safe area.

OVERHEAD TRANSPARENCIES

Masters for overhead transparencies are found on the following pages. Transparencies can be made by using clear acetate sheets in a photocopy machine. This can be done at any commercial copy shop, or by following this method on your office machine:

1. Use standard 8½ × 11 clear acetate sheets, which can be obtained at office supply stores, stationery stores, or commercial copying centers.
2. Check the photocopier's manual to make sure it will take the thicker acetate sheets.
3. Remove the paper supply in the photocopying machine and replace it with a supply of acetate sheets.
4. Make photocopies from the masters on the following pages in the usual way.
   The material will be reproduced on the clear acetate sheets for use on any overhead projector.

TRANSPARENCY 1

TASK MODULE 23109, INTRODUCTION TO TROUBLESHOOTING

COURSE OBJECTIVES

Upon completion of this module, the trainee will be able to:

1. **Describe a basic troubleshooting methodology that can be used to identify a variety of concrete construction problems and their causes.**

2. **Identify problems with fresh concrete and describe ways to prevent them.**

3. **Identify different concrete defects such as crazing, cracking, dusting, scaling, popouts, and efflorescence, and describe ways to prevent them.**

TRANSPARENCY 2

TASK MODULE 23109, INTRODUCTION TO TROUBLESHOOTING

PERFORMANCE PROFILE TASKS

1. Identify from a set of photographs the following problems:
   - Dusting
   - Crazing
   - Spalling
   - Popouts
   - Honeycombs
   - Scaling
   - Plastic shrinkage cracking
   - Excess bleeding
   - Drying-shrinkage cracks
   - Bugholes

TRANSPARENCY 3

TASK MODULE 23109, INTRODUCTION TO TROUBLESHOOTING

SAFETY REQUIREMENTS

- Wear Appropriate Personal Protective Equipment
- Obey all safety rules and procedures
- Be aware of potential work site hazards and specific safety requirements
- Pay attention to all environmental safety requirements

TRANSPARENCY 4

FIGURE 1

TASK MODULE 23109, INTRODUCTION TO TROUBLESHOOTING

TROUBLESHOOTING METHODOLOGY

109F01.EPS

TRANSPARENCY 5

TASK MODULE 23109, INTRODUCTION TO TROUBLESHOOTING

CORRECTIVE ACTION

Fix the cause of the problem:

- As well as its symptoms
- As quickly and economically as possible
- Without creating any new problems

NOTE ON PERFORMANCE PROFILE TESTING

Performance Profiles are included in this Instructor's Guide in a format that can be easily photocopied for each trainee. The Profiles measure trainee competency in the tasks taught in this module.

Please note the number of tasks to be tested while teaching this module. Each trainee should be tested on the tasks listed in the Performance Profile. Before the performance testing, the instructor should brief the trainees on:

- test objectives and criteria,
- safety precautions, and
- procedures for each task to be tested.

The instructor administering the performance testing should also do the following:

- ensure that all of the needed equipment is available and operating properly;
- set up the testing stations;
- organize and administer the test in a way that allows for optimal performance;
- complete the Performance Profile Sheet for each trainee by assigning a score for each listed task;
- monitor adherence to all safety regulations and precautions;
- provide adequate supervision to prevent injuries; and
- take immediate and effective action to remedy any emergency.

ACCREDITATION TESTING

If this Performance Profile Testing is done in the National Center for Construction Education and Research Standardized Craft Training Program, the following conditions must be met:

1. The Craft Instructor must hold valid NCCER instructor certification.
2. The training must be delivered through a Training Program Sponsor recognized by the NCCER.
3. For every module, the specific performance testing must be completed to the satisfaction of the instructor.
4. The results of the testing must be recorded on the Craft Training Report Form. This form must be provided to the local Training Program Sponsor to be forwarded to the NCCER National Craft Training Registry.

Craft: Concrete Finishing	WHEELS
Task Module Number: 23109	OF
Task Module Title: Introduction to Troubleshooting	LEARNING

TRAINEE NAME: ______________________________

TRAINEE SOCIAL SECURITY NUMBER: ______________________________

CLASS: ______________________________

TRAINING PROGRAM SPONSOR: ______________________________

____

INSTRUCTOR: ______________________________

Rating Levels:
1. Passed: performed task.
2. Failed: did not perform task.

Recognition: When testing for the NCCER Standardized Craft Training Program, be sure to record Performance Profile testing results on Craft Training Report Form 200 and submit the results to the Training Program Sponsor.

Objective	TASK	RATING
2 & 3	Identify 10 different types of concrete problems from photographs.	

109F03.TIF

1. ____________________________

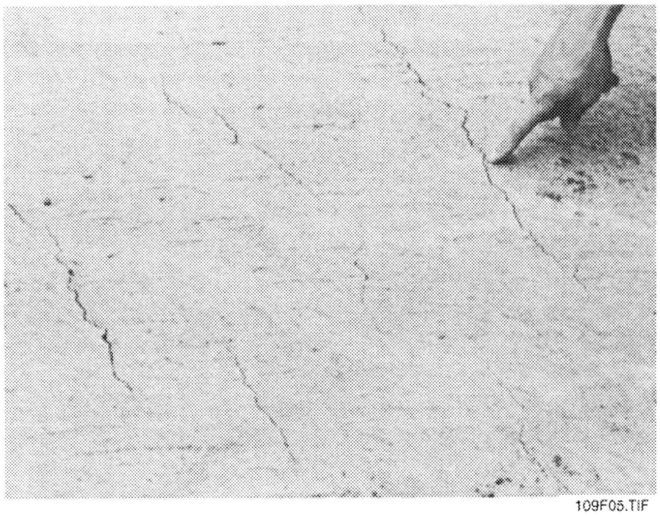

109F05.TIF

2. ____________________________

109F09.TIF

3. ______________________

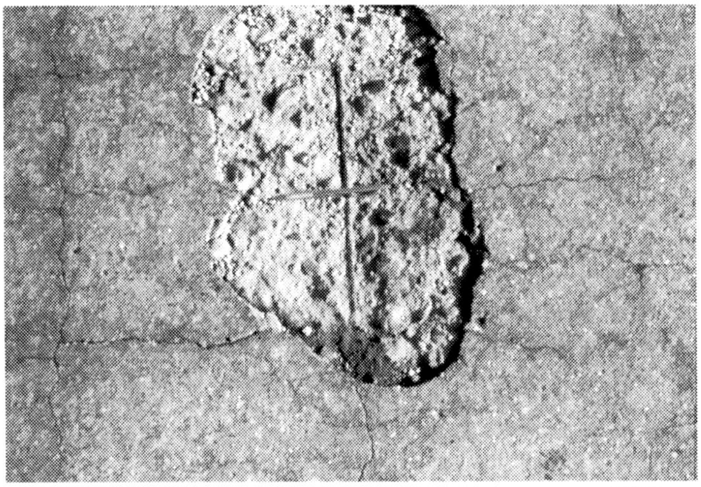

109F11.TIF

4. ______________________

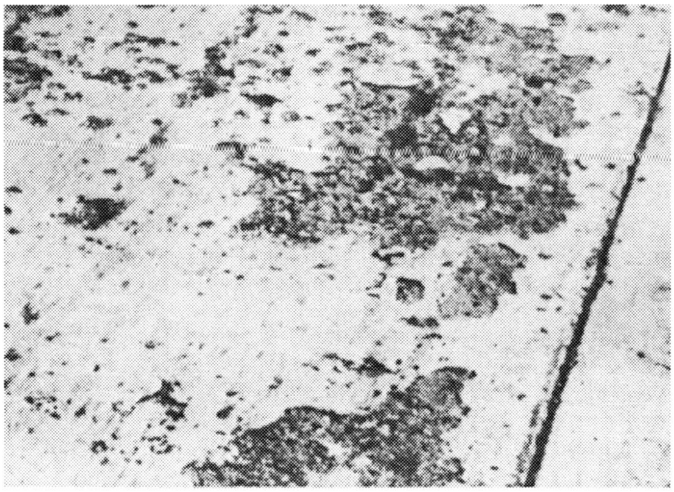

109F10.TIF

5. ______________________________

109F07.TIF

6. ______________________________

109F12.TIF

7. ______________________________

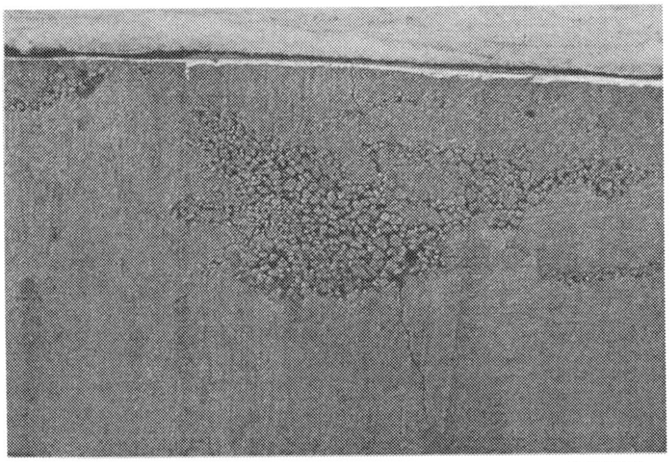

109F16.TIF

8. ______________________________

109F17.TIF

9. ____________________________

109F08.TIF

10. ____________________________

Performance Profile Examination Answer Sheet

1. Excess bleeding
2. Plastic-shrinkage cracks
3. Dusting
4. Spalling
5. Scaling
6. Crazing
7. Popouts
8. Honeycombs
9. Bugholes
10. Drying-shrinkage cracks

NCCER CRAFT TRAINING USER UPDATES

The NCCER makes every effort to keep these manuals up-to-date and free of technical errors. We appreciate your help in this process. If you have an idea for improving this manual, or if you find an error, a typographical mistake, or an inaccuracy in the NCCER's Craft Training Manuals, please write us, using this form or a photocopy. Be sure to include the exact module number, page number, a description of the problem, and the correction, if possible. Your input will be brought to the attention of the Technical Review Committee. Thank you for your assistance.

Instructors – If you found that additional materials were necessary in order to teach this module effectively, please let us know so that we may include them in the Equipment/Materials list in the Instructor's Guide.

Write: Curriculum and Revision Department
National Center for Construction Education and Research
P.O. Box 141104
Gainesville, FL 32614-1104

Fax: 352-334-0932

Craft ______________________ Module Name ______________________

Module Number ______________________ Page Number(s) ______________________

Description of Problem

____

____

____

____

(Optional) Correction of Problem

____

____

____

(Optional) Your Name and Address

____

____

____

notes